# TIME

## *Apprehend the Future by Analyzing Time by Calculation*

ABDOUL KHOUDOUSS KAMARA

ISBN 979-8-89130-440-6 (paperback)
ISBN 979-8-89130-441-3 (digital)



Christian Faith Publishing
832 Park Avenue
Meadville, PA 16335
www.christianfaithpublishing.com

Printed in the United States of America

# Contents

# Preface

The periodicity of the natural phenomena of the environment in which man lives has led him to create a calendar that is a dating according to time. Observation of the cycles of the seasons and lunar and solar movements are the first clues to ancient calendars. The latter are the organizers of religious, social, and agricultural events in those distant times.

There are different types of calendars, the best known of which is the Gregorian. They can be qualified as solar or lunar depending on whether they are established on the movements of the sun or the moon.

Our study focuses on the Gregorian calendar. In everything that follows, we will only use the latter for our calculations.

In some human communities, certain dates are not recommended for doing business, traveling, or any other activity. Those interested in mysticism or in demonstrating these beliefs can make use of these periodic calendars which can be of great help to them.

Are natural phenomena linked by a periodicity in time?

Can we predict the future by calculating calendar intervals?

We can predetermine or predict the future through complex temporal relationships that link seemingly independent natural phenomena.

Two events occurring on separate dates may seem hazardous as they are united by calendar periodicity as we attempt to demonstrate in this book.

Our goal is that, in the near future, mankind can create, using the mathematical formulas in the book, software algorithms for predicting these events.

The scientific community can use the work of this book to improve knowledge of anticipation: calculating the future or the engineering of time.

The particular interest of this book is the study of the periodicity of natural phenomena about the periodicity of calendars or the study of the correlations between phenomena about time.

We can, by using this study, make a link between the years of good floods for the wines. We could do it in other areas: years of economic growth, stock market crashes, weather, war, criminology, the fight against terrorism, SARS disease, agricultural production, etc.

This study may also be of interest to anyone whose work is related to time. Thus, the judge, the security services, social security, meteorologists, farmers, computer scientists, for the study of prediction algorithms using Big Data databases, and any scientist in general, can benefit from this book.

Are there more warring years than others? The answer to this question is at the very heart of this book. That is, studying the relationship between facts and time.

Other questions in the astrological field are as follows: Are there affinities between people of different calendars? Are there favorable days or years to start a business or undertake any activity? We make calendars and their renewals available to the reader, and everyone can deduce the phenomena they observe while referring to the calendars and their periodicity.

Historical archival work about comparisons of calendar intervals can help us take precautions for certain years to come.

We start from the assumption that chance does not exist and that phenomena are linked by hidden temporal relationships. Through further research, we can discover these relationships.

This is the purpose of the development of timetables so that the researcher can have in his hands a tool for comparison.

Working with databases, artificial intelligence coupled with supercomputers and Internet search engines can help us predict phenomena, that is, the use of Big Data for the prediction of future phenomena.

The aim is to pay attention to certain years by studying previous years belonging to the same calendar periods.

# Introduction

Since ancient times, remarkable minds have held reflections on time. Chronologically we will start with the Buddhists. Ivan Mahayana showed the hadronic constitution of matter. All the elements of the universe are related by time.

## What Is Time?

In the seventh century, the prophet Muhammad, PBUH, told his disciples not to criticize time because time is God.

The modern scientists of the twentieth century arrived at conclusions not very different from those of the religious ones.

Let's talk about John Archibald Wheeler first; he finds that time is made up of corpuscles called Geons.

By virtue of the universal principle of symmetry, Geons have their anti-Geons, which are the building blocks of anti-time.

Soviet scientist Nikolai Kozyrev considers time a form of energy. Energy is not only linked to the physical phenomenon but also to the phenomenon studied by parapsychology. Kozyrev says time is the most important and most enigmatic element in the universe.

It does not propagate like light waves and manifests everywhere at once and instantly. Time is also present; it is time that connects us to others and connects everything in the universe. Are these not properties of immanence and ubiquity?

Professor Fritjof Capra comes to the same conclusions as Kozyrev, a Soviet scientist.

I started from a formula given to me by Diop Alassane from the village of Maghama, Mauritania. What interested me first was knowing how to calculate the day of the week that corresponds to a given date. Then I was interested in finding periodicities. Gradually

I discovered that the Gregorian calendar actually consisted of fourteen calendars, seven leap and seven non-leap years. By observing the spectra of calendars, I was able to demonstrate the relationships that existed between leap and non-leap calendars. I determined the renewal of calendars by studying what I called the ambulation of each calendar.

So I found a rule to simplify the calculations. For example, for calendar 24, a date between March 1 and December 31, we can calculate with calendar 2. They are identical for this time interval. The same goes for a date between January 1 and February 29 for calendars 20 and 3. By studying the book, we can simplify the calculations by studying the relationships between the calendars 20/3 and 24/2 then the others, 12/1 and 5/16, 6/12, etc. (See the time diagrams on the Appendix page.)

## People, Organizations, and Institutions
## Targeted for the Use of Calendars

Anyone may want to know the day that corresponds to their date of birth. Followers of the esoteric sciences often need to know the date of birth.

Apart from these particular cases, the police, the courts, and the social security establishments need to know the day of the week corresponding to a given date to reveal the truth or to establish statistics.

Social security agents need to know the date of an accident that caused a sick leave and the date of resumption of the old accident a few years ago to establish a statistic.

I was able to calculate the indices for the months of other centuries using the formula given to me by my friend Alassane Diop, who said, "Mustache for close friends." So then I discovered that there was a four-hundred-year periodicity between the centuries. So I can tell you that the centuries 1201 to 1300 and the century 1601 to 1700 are identical. 2001 to 2100 reproduces identically to previous centuries: the month indices are identical. And the centuries 1501 to 1600 and 1901 to 2000 are reproduced identically. Their month indices are identical.

## Calendar Properties

We saw above that scientists agree that everything in the universe is time-bound. No element is isolated.

Periodicity means similarity. Consequently, this similarity will not remain without consequence. This is why we really think that the entities we call calendars have certain similarities. What the Chinese certainly think—the year of the rats, the dogs, the elephants, etc.—the Chinese have twelve renewing years while I have found fourteen different calendars. By giving names to the years, snakes, tigers, and buffaloes, we notice that the Chinese years are twelve in number while our calendars are fourteen periodicals.

Now we come to the most interesting part of the study, that is, how we can use these calendars and determine their main properties.

Indeed, a calendar can have multiple properties—years of good harvests, natural disasters, and armed conflicts between nations. For a given calendar, the properties are only valid for one century. This is why the days of the week must also be taken into account. For that, it is necessary to study well the diagrams of time, diagrams which we gave and included in this study. Taking all these elements into account, one can decipher and decipher everything that is in the archives.

Only observations with a frequency of at least 70 percent will be considered significant. Everything we have said is valid globally and not just locally.

## The Investigations Necessary for This Document to Be Scientific

They must be led by people whose sagacity is foolproof. They must be carried out in all parts of the world and by several teams.

We know that 2013 is a repeat of 2002 which is the year of the US invasion of Afghanistan. In 2013, armed conflicts in West Africa were numerous. There were conflict in Syria and threats between China and Japan and threats of war between the two Koreas in 2013. In 2005, the city of New Orleans was destroyed by torrential rains and the exodus of residents. It was Hurricane Catharina. In 2011, which is a repeat of 2005, Fukushima was destroyed by a vast tsunami. These two examples do not sufficiently characterize calendars 2 and 5. Indeed, a swallow does not make the spring. This is why we asked for teams that scrutinize all the archives of the world for the trends that emerge and more comprehensive and reliable statistics.

We are awaiting information concerning the years 2019 and 2022 which are the repetitions of calendars 2 and 5. Through these investigations, we must be attentive to the smallest elements that may be moderating or that lead to exasperation or paroxysm.

## Periodicity or DNA of Time

Ultimately, I am happy to have made this great discovery, namely the discovery of the fourteen calendars and periodicities which could be the DNA of time.

Fiery impulses might have prompted me to call these fourteen calendars the genome of time.

But the caution dictated by the high probability of existence and the considerable influence of parallel universes on our world has tempered my enthusiasm.

Finally, I am happy to have found the demonstration that in ten thousand years, the Gregorian calendar which we use will be three days behind reality. This is what my history teacher told me in high school but without demonstration.

## Our Mathematical Mission

The formula that we used and that we qualified as empirical (I relate below the circumstances during which this formula was communicated to me) is surely a cousin of the formula of Christian Zeller established by him at the end of the nineteenth century, in the '80s. Using this cousin formula, the calculations revealed that there are fourteen Gregorian calendars (seven leap calendars and seven non-leap calendars) which are always renewed, all the time. These calendars have constant and fixed relationships with each other; these calculations revealed the periodicity of each calendar. Furthermore, these calculations have shown that there is a narrowed periodicity (four hundred years) and a general periodicity of ten thousand years.

During July 1972, in my living room in Nouakchott, where my friends and I were gathered around a tea, one of my comrades announced to us that he had a formula by which he could determine the day of the week corresponding to a given date.

Our attention was keenly drawn, and we wanted to test his formula and ask him for the day of the week that corresponded to a given date. What day of the week corresponded to July 17, 1972? He told us we had to know the formula literally. Here is the formula:

Date + index of the month + year + NLY; the whole divided by $7 = K + (R / 7)$

NLY is the number of leap years.

Each month has a clue; the NLY is the whole part of the year divided by 4.

$R$ is the row of the day of the week.

$K$ is the integer part of the division by 7.

Let us calculate the day of the week which corresponds to the date of July 17, 1972; the index for July is 6.

$$(17 + 6 + 72 + 18) / 7 = (16 + 1/7)$$

The remainder of the division of $113/7 = 1$.

1 is the rank of the day sought; it is a Monday.

Let us give the value of all the indices of the months:

- January: 0
- February: 3
- March: 3
- April: 6
- May: 1
- June: 4
- July: 6
- August: 2

- September: 5
- October: 0
- November: 3
- December: 5

The above indices are valid from 1901 to the year 2000. Actually, the friend had just given us two or three indices, and the others I established by testing the formula using a calendar that I had in front of me.

Subsequently, I established indices for the other centuries. During the study, it occurred to me that time is formed by what I call calendars, that these calendars have relationships with each other, and that they have a certain periodicity. At the end of the study, we will try to say how reliable the calendars are over time.

If the remainder of the division by 7 is 0, it is Sunday.

If the rest is 1, it's Monday.

If the rest is 2, it's Tuesday.

If the rest is 3, it's Wednesday.

If the remainder is 4, it's Thursday.

If the remainder is 5, it's Friday.

If the rest is 6, it's Saturday.

# Time Structure

## Periodicity of Leap Years

Q = Date I = Index of the month

M = Year R = Rest of the division by 7

E (M / 4) read whole part of the year divided by 4.

(M / 4) = NLY = number of leap years since 1901.

Date: when we say that we are July 5, 1912, 5 is the date, the index of the month is 6, the year is 12, and the NLY is the number of leap years since 1901. The whole part of $(M/4)$ is denoted $E(M/4)$.

The calculation is therefore done as follows:

$$\frac{Q + I + M + E(M/4)}{7} = K + (R/7)$$

$K$ is an integer varying from 0 to plus infinity.
$E(M/4)$ is the integer part of $(M/4)$.
Example:
$E(37/4) = 9$, instead of 9.25.
Consider two years, $M1$ and $M2$, whose dates, $Q1$ and $Q2$, are identical, that is to say, which occur on the same day of the week. So it is not necessarily true that the time that elapses between $M1$ and $M2$ is the period of the calendar.

As in the example, in the following formula:

$$I1 = I2$$
$$Q1 = Q2$$

However, $M1$ is different from $M2$.

$$\frac{(Q + I1 + M1 + E\,(M1\,/4)}{7} - \frac{(Q2 + I2 + M2 + E\,(M2\,/\,4)}{7}$$
$$= K1 - K2 + (R1\,/\,7) - (R2\,/\,7)$$

Indeed, the two remainders, $R1$ and $R2$, are equal since it is the row of the day of the week.

So $(R1\,/\,7) - (R2\,/\,7) = 0$.

Looking at the calendars that I kept for years made me think about periodicity.

The question is, can these calendars not be used for years to come?

$$\frac{Q1 + I1 + M1 + E\,(M1\,/\,4)}{7} - \frac{Q2 + I2 + M2 + E\,(M2/4)}{7} = K1 - K2$$

$$(M1 + E\,(M1\,/\,4)) - (M2 + E\,(M2\,/\,4)) = 7\,(K1 - K2)$$

If $M1$ and $M2$ are leap, $E\,(M1\,/\,4) = (M1\,/\,4)$ and $E\,(M2\,/\,4) = (M2\,/\,4)$

So we can write:

$$(M1 - M2) + (E\,(M1\,/\,4)) - (E\,(M2/4)) = (M1 - M2) + ((M1 - M2)\,/\,4)$$
$$= \frac{4\,(M1 - M2) + M1 - M2}{4}$$

$$\frac{5M1 - 5M2}{4} = 7\,(K1 - K2)$$

$$(5/4)\,(M1 - M2) = 7\,(K1 - K2)$$

Hence, $M1 - M2 = \dfrac{((7x4) \, x \, (K1 - K))}{5} = \dfrac{28K}{5}$

$\acute{K} = K1 - K2$, If $\acute{K} = 5$ then $M1 - M2 = 28$

We have just established that the great periodicity is equal to twenty-eight years.

# The Calendars

We have established that there is a periodicity in time when it comes to leap calendars. Is the timeline unique, or are we dealing with multiple entities that together make up the timeline as we know it? In year 1, how does it renew itself periodically? How are non-leap years renewed? The purpose of this chapter is to answer these different questions.

## Calendar 1

Let's take the empirical formula at the base of our entire study:

$$\frac{Q + I + M + E\,(M\,/\,4)}{7} = K + (R\,/\,7)$$

Let's find out the day of the week that corresponds to the date of January 1, 1901:

$$\frac{1 + 0 + 1 + 0}{7} = 0 + (2/7)$$

So it's a Tuesday (the second day of the week). The years 1901 to 2000 which January 1 is a Tuesday are the renewal of year 1, calendar 1. We are going to list all these years.

$$\frac{1 + 0 + M + E\,(M\,/\,4)}{7} = K + (2/7)$$

$$1 + 0 + M + E\,(M\,/\,4) = 7K + 2$$

$$M + E\,(M\,/\,4) = 7K + 2 - 1$$

$$M + E\,(M\,/\,4) = 7K + 1$$

($M$ + $E$ ($M$ / 4)) congruent to 1 modulo 7.

($M$ + $E$ ($M$ / 4)) belongs to the list $N$ of numbers which are congruent with 1. But $M$ + $E$ ($M$ / 4) is identical only to part of this list. For example, if $N$ = 15, there is no $M$ which satisfies the equation. The equation $M$ + $E$ ($M$ / 4) = 15 has no solution. This equation has no solution.

So $N$ = 15 is to be eliminated from the list of the series of numbers. The same goes for other numbers in the series 29.50, 64, etc. You can see the calculations below. Congruence at 1 is the different values of ($M$ + $E$ ($M$ / 4)) divided by 7 whose remainders are always equal to 1.

Let us draw the set of numbers $N$ which is congruent with 1 modulo 7. This set, let us call it "spectrum set of 1."

$$N = M + E (M/4)$$

| $N$ | = |
| --- | --- |
| 1 | 71 |
| 8 | 78 |
| 15 | 85 |
| 22 | 92 |
| 29 | 99 |
| 36 | 106 |
| 43 | 113 |
| 50 | 120 |
| 57 | 127 |
| 64 | 134 |

The years 1901, 1907, 1918, 1929, 1935, 1946, 1957, 1963, 1974, 1985, 1991, and 2002 are the renewal of calendar 1. They are represented by absolutely identical calendars, from January 1 to December 31 on the days of the week, and the dates correspond exactly to each one. We call these years the nodal points of the calendar spectrum 1. $M$ = 1, 7, 18, 29, 35, 46, 57, etc. are the years of the years 1901, 1907, 1918, 1929, 1935, 1946, and 1957, etc.

The values of $M$ are given by the calculations below. $M$ is calculated from the list $N$ of numbers which congruent with 1 modulo 7.

$$M + E (M / 4) = N; \quad N \text{ belongs to the set above.}$$

If $M + E (M / 4) = N = 1 => M = 1 (1 + E (1/4)) = 1$

$M + E (M / 4) = N = 8$ so $M = 7 (7 + E (7/4)) = 7 + 1 = 8$; therefore, $M = 7$ which corresponds to 1907 is the renewal of calendar 1.

If $N = 22$, then $M = 18$ because $(18 + E (18/4)) = 18 + 4 = 22$; therefore $M = 18$ which corresponds to 1918 is the renewal of calendar 1

If $N = 36$, then $M = 29$ because $(18 + E (29/4)) = 29 + 7 = 36$; therefore 29 is the renewal of calendar 1.

If $N = 43$, then $M = 35$ because $(35 + E (35/4)) = 35 + 8 = 43$, so 35 is the renewal of calendar 1.

If $N = 57$, then $M = 46$ because $(46 + E (46/4)) = 46 + 11 = 57$, so 46 is the renewal of schedule 1.

If $N = 71$, then $M = 57$ because $(57 + E (57/4)) = 57 + 14 = 71$, so 57 is the renewal of calendar 1.

If $N = 78$, then $M = 63$ because $(63 + E (63/4)) = 63 + 15 = 78$, so 63 is the renewal of calendar 1.

If $N = 92$, then $M = 74$ because $(74 + E (74/4)) = 74 + 18 = 92$, so 74 is the renewal of schedule 1.

If $N = 106$, then $M = 85$ because $(85 + E (85/4)) = 85 + 21 = 106$, so 85 is the renewal of calendar 1.

If $N = 113$, then $M = 91$ because $(91 + E (91/4)) = 91 + 22 = 113$, so 91 is the renewal of calendar 1.

If $N = 127$; therefore $M = 102$ because $(102 + E (102/4)) = 102 + 25 = 127$; therefore 102 corresponds to the year 2002 is the renewal of calendar 1.

Let's try to see what the structure of this new set is as follows: 1, 7, 18, 29, 35, 46, 57, 63, 74, 85, 91, 2002.

$$7 - 1 = 6 \quad 35 - 29 = 6 \quad 63 - 57 = 6$$
$$18 - 7 = 11 \quad 46 - 35 = 11 \quad 74 - 63 = 11$$
$$29 - 18 = 11 \quad 57 - 46 = 11 \quad 85 - 74 = 11$$

We can continue and symmetrical elements appear in this table. First of all:

$$29 - 1 = 28$$
$$57 - 29 = 28$$
$$85 - 57 = 28$$

The twenty-eight-year period appears. Then we notice that during the twenty-eight-year period for a calendar, the reproduction of the calendar is done as follows:

The first reproduction occurs six years after the first year; the second reproduction occurs eleven years after the seventh year; the third reproduction occurs eleven years after the eighteenth year.

The reproduction of calendar 1 is therefore characterized by the numbers 6, 11, and 11 during the first period of twenty-eight years. Numbers 6, 11, and 11 are the pseudo-periods or Harmonics of the twenty-eight-year period. Let's call the pseudo-periods the ambulation of calendar 1

The years 1, 7, 18, 29, 35, 46, 57, etc. are absolutely identical and are the reproduction of the same calendar. If we take any date, the day of the week corresponding to this date is the same for all the aforementioned calendars.

## Calendar 2

Let's go back to the empirical formula and find the day of the week that corresponds to January 1, 1902.

$$\frac{Q + I + M + E\,(M\,/\,4)}{7} = K + (R\,/\,7)$$

$$\frac{1 + 0 + 2 + 0}{7} = 0 + (3/7)$$

So it's a Wednesday. Let's find out what are the years, between 1901 and 2000, in which January 1 is a Wednesday.

$$\frac{1 + 0 + M + E\,(M\,/\,4)}{7} = K + (3/7)$$
$$M + E\,(M\,/\,4) = 7K + (3 - 1)$$
$$M + E\,(M\,/\,4) = 7K + 2$$

So $(M + E\,(M\,/\,4))$ congruent with 2 modulo 7. $(M + E\,(M\,/\,4))$ belongs to the series of numbers $N$ which is congruent with 2 modulo 7.

But as before, $(M + E\,(M\,/\,4))$ is only identical to a part of this series which satisfies the equality. For example, if $N = 9$, there is no $M$ which satisfies the equation, the equation $M + E\,(M/\,4) = 9$ has no solution. This equation has no solution. So $N = 9$ is to be eliminated from the list of the series of numbers. The same goes for other numbers in the series 30, 44, 65, etc. You can see the calculations below.

$M$ = 2, 13, 19, 30, 41, 47.58, etc. are the years of the years 1902, 19013, 1919, 1930, 1941, 1947, 1958, etc.

The values of $M$ are given by the calculations below. $M$ is calculated from the list $N$ of the numbers which congruent with 2 modulo 7.

Let us list the numbers $N$ which are congruent with 2 modulo 7.

- 2, 51, 100 (2000)
- 9, 58, 107
- 16, 65, 114
- 23, 72, 121
- 30, 79, 128
- 37, 86, 135
- 44, 93, 142 (2042)

If $N = 2$, then $M = 2$ because $2 + E\,(2/4) = 2 + 0 = 2$.

If $N = 16$, then $M = 13$ because $13 + E\,(13/4) = 13 + 3 = 16$.
If $N = 23$, then $M = 19$ because $19 + E\,(19/4) = 19 + 4 = 23$.
If $N = 37$, so $M = 30$ because $30 + E\,(30/4) = 30 + 7 = 37$.
If $N = 51$, then $M = 41$ because $41 + E\,(41/4) = 41 + 10 = 51$.
If $N = 58$, then $M = 47$ because $47 + E\,(47/4) = 47 + 11 = 58$.
If $N = 72$, then $M = 58$.
If $N = 86$, then $M = 69$.
If $N = 93$, so $M = 75$.
If $N = 107$, then $M = 86$.
If $N = 121$, then $M = 97$.

The years 1902, 1913, 1919, 1930, 1941, 1947, 1958, 1969, 1975, 1986, and 1997 are the renewal years of calendar 2. They are represented by absolutely identical calendars, from January 1 to December 31, the days of the week and the dates correspond exactly to each one.

Let's try to see what the structure of this new set is as follows: 2, 13, 19, 30, 41, 47, 58, 69, 75, 86, 97.

$$13 - 2 = 11; 41 - 30 = 11; 69 - 58 = 11$$
$$19 - 13 = 6; 47 - 41 = 6; 75 - 69 = 6$$
$$30 - 19 = 11; 58 - 47 = 11; 86 - 75 = 11$$

We can continue and symmetrical elements appear from this table.

First: $30-2 = 28$, $58-30 = 28$, and $86-58 = 28$. The 28-year period then appears. We notice that during the period of 28, the reproduction of calendar two from 1901 to 2000 is as follows.

The first reproduction occurs eleven years after the first year and then the second reproduction occurs six years after the thirteenth year. The third reproduction occurs eleven years after the nineteenth year. That is, twenty-eight years after 1902. The calendar is therefore characterized by the numbers 11, 6, and 11 over twenty-eight years. 11, 6, and 11 are the pseudo-periods or Harmonics of the twenty-eight -year period. The ambulation of calendar 2 is therefore 11, 6, and 11.

## Calendar 3

Let's go back to the empirical formula and find the day of the week that corresponds to January 1, 1903.

$$\frac{1 + 0 + 3 + 0}{7} = 0 + (4/7)$$

So it's a Thursday. Let's find out which years January 1 is a Thursday.

$$\frac{1 + 0 + M + E\,(M\,/\,4)}{7} = 0 + (4/7)$$
$$M + E\,(M\,/\,4) = 7K + (4 - 1)$$
$$M + E\,(M\,/\,4) = 7K + 3$$

$M + E\,(M\,/\,4)$ congruent with 3 modulo 7; $M + E\,(M\,/\,4)$ matches the years $N$ with numbers which is congruent with 3 modulo 7, but as before, again, $M + E\,(M\,/\,4)$ is only identical to a part of this series which satisfies the equality:

$$M + E\,(M\,/\,4) = N.$$

Let us list the numbers $N$ which are congruent with 3 modulo 7.

| | | |
|---|---|---|
| 3 | 52 | 101 |
| 10 | 59 | 108 |
| 17 | 66 | 115 |
| 24 | 73 | 122 |
| 31 | 80 | 129 |
| 38 | 87 | 136 |
| 45 | 94 | 143 |

With $M + E\,(M/4) = N$:

If $N = 3$, then $M = 3$.
If $N = 17$, then $M = 14$.
If $N = 31$, so $M = 25$.
If $N = 38$, so $M = 31$.
If $N = 52$, then $M = 42$.
If $N = 66$, then $M = 53$.
If $N = 73$, then $M = 59$.
If $N = 87$, so $M = 70$.
If $N = 101$, then $M = 81$.
If $N = 108$, so $M = 87$.
If $N = 122$, then $M = 98$ (1998).

The years 1903, 1914, 1925, 1931, 1942, 1953, 1959, 1970, 1981, 1987, and 1998 are represented by absolutely identical calendars, from January 1 to 31 December the days of the week and the dates correspond exactly to each one.

Let's try to see what is the structure of this new set and the relations that characterize it.

$$(3, 14, 25, 31, 42, 53, 59, 70, 81, 87, 98)$$

$$14 - 3 = 11;\ 42 - 31 = 11;\ 70 - 59 = 11;\ 98 - 87 = 11$$
$$25 - 14 = 11;\ 53 - 42 = 11;\ 81 - 70 = 11$$
$$31 - 25 = 6;\ 59 - 53 = 6;\ 87 - 81 = 6$$

We can continue and symmetrical elements appear in this table. First of all: $31 - 3 = 28,\ 59 - 31 = 28,\ 87 - 59 = 28$.

The twenty-eight-year period appears, then we notice that during the twenty-eight-year period, the reproduction of calendar three from 1901 to 2000 is as follows:

The first reproduction occurs eleven years after the third year, the second reproduction occurs eleven years after the fourteenth year, and the third reproduction occurs six years after the twenty-fifth year, that is, twenty-eight years after 1903.

Calendar 3 is therefore characterized by the numbers 11, 11, and 6 over a period of twenty-eight years. 11, 11, and 6 are the pseudo-periods or harmonics of the twenty-eight-year period. The ambulation of calendar 3 is therefore 11, 11, 6.

More generally, a calendar that is located after year 0 or a leap year has an ambulation similar to that of calendar 1, that is to say, 6, 11, 11, for the period of twenty-eight years. So 1941 comes until after 1940, and the latter is a leap, so the reproduction of 1941 will be done according to the ambulation 6, 11, and 11 for a period of twenty-eight years. The reproduction of 1941 is carried out by the calendars of the following years: 1941, 1947, 1958, 1969. A calendar which is located two years after the year 0 or two years after the first leap year has an ambulation as follows: 11, 6, 11. Thus 1942 is reproduced by the following calendars: 1942, 1953, 1959, 1970.

The calendar which is located three years after the year 0 or 3 years after the first leap year has an ambulation of 11, 11, 6. Thus the reproduction of 1943 is carried out in the following years: 1943, 1954, 1965, 1971.

## Calendar 4

Let's go back to the empirical formula and find the day of the week that corresponds to January 1, 1904.

Calendar 4-1 from January 1 to February 29

This calendar 4 is leap and has two parts: the first part from January 1 to February 29 and the second part from March 1 to December 31.

For calendar 4 from January 1 to February 29, the formula must be applied with caution. In fact, the year is only leap from March 1.

So from January 1 to February 29:

$$E \left( M / 4 \right) = E \left( (M - 1) / 4 \right) = E \left( 3/4 \right) = 0$$

Using the empirical formula, we can find the day that corresponds to the date of January 1, 1904.

$$\frac{1 + 0 + 4 + E\left((4-1)/4\right)}{7} = 0 + (5/7)$$

So it's a Friday. The years 1901 to 2000 where January 1 is a Friday are the renewal of the 4-1 calendar. We will establish the list of congruent numbers with 4 modulo 7 or the spectrum of the calendar 4-1:

$$\frac{1 + 0 + M + E\left((M-1)/4\right)}{7} = K + (5/7)$$

$$\left(M + E\left((M-1)/4\right)\right) = 7K + 5 - 1,$$
$$\text{so } \left(M + E\left((M-1)/4\right)\right) = 7K + 4$$

$\left(M + E\left((M-1)/4\right)\right)$ is congruent with 4 modulo 7. Let us draw the set of the elements of the spectrum 4-1.

| | | |
|---|---|---|
| 4 | 46 | 88 |
| 11 | 53 | 98 |
| 18 | 60 | 102 |
| 25 | 67 | 109 |
| 32 | 74 | 116 |
| 39 | 81 | |

If $\left(M + E\left((M-1)/4\right)\right) = N = 4n = 4$, the first node is $M = 4$. Like $N = 4n$, $N$ is a multiple of 4 because that $M$ is a leap. $M$ must meet the following conditions:

(a) $M$ is a multiple of 4 (leap year)
(b) The sum of $\left(M + E\left((M-1)/4\right)\right)$ must have $4n$ of the spectrum elements as the image.

If $M + E((M-1)/4) = N = 39$, which gives $M = 32$.

$$N = 39; \quad 32 + E(31/4) = 32 + 7 = 39$$
$$N = 74; \quad 60 + E(59/4) = 60 + 14 = 74$$
$$N = 109; \quad 88 + E(87/4) = 88 + 21 = 109$$

The years 4, 32, 60, and 88 are the same as the calendar 4-1; the ambulation of 4, 32, and 60 is twenty-eight years. $4 + 28 = 32$; $32 + 28 = 60$; $60 + 28 = 88$. The periodicity is twenty-eight years.

The 4-2 calendar from March 1 to December 31

Let's find the day of the week that corresponds to March 1, 1904.

$$\frac{1 + 3 + 4 + E(4s/4)}{7} = \frac{1 + 3 + 4 + 1 = 9/7}{7} = 1 + 2/7$$

So it's a Tuesday, and we'll be building the spectrum for the 4-2 schedule.

$$\frac{1 + 3 + M + E(M/4)}{7} = K + (2/7)$$

$M + E(M/4) = 7(K-1) + (7 + 2 - 4) = 7K' + 5K - 1 = K'$ is also an integer.

$M + E(M/4)$ is congruent with 5 modulo 7. We will establish the spectrum of calendar 4 from March 1 to December 31.

| | | |
|---|---|---|
| 5 | 47 | 89 |
| 12 | 54 | 96 |
| 19 | 61 | 103 |
| 26 | 68 | 110 |
| 33 | 75 | 117 |
| 40 | 82 | 124 |
| | | 131 |

The nodes are as follows:

$M + E\,(M\,/\,4) = 5 = N$, therefore $M = 4$; if $M = 4$, so $4 + E\,(4/4)$ $= 5 = N$.

If $N = 40$, then $M = 32$ because $32 + (32/4) = 32 + 8 = 40$.

If $N = 75$, then $M = 60$; $60 + (60/4) = 60 + 15 = 75$.

If $N = 110$, then $M = 88$; $88 + (88/4) = 88 + 22 = 110$.

First of all, we recognize the periodicity of twenty-eight years: $32 - 4 = 28$; $60 - 32 = 28$; $88 - 60 = 28$.

Years 4, 32, 60, and 88 are identical to the 1904 calendar from March 1 to December 31. The ambulation is twenty-eight years, and the periodicity is twenty-eight years.

## Calendar 5

Let's apply the empirical formula to calculate the day of the week that corresponds to the date of January 1, 1905.

$$\frac{1 + 0 + 5 + E\,(5/4)}{7} = \frac{1 + 0 + 5 + 1}{7} = 7/7 = 1 + 0/7$$

So it's a Sunday. The years in which January 1 is a Sunday are the renewal of calendar 5. We will therefore establish the spectrum of these years.

$$\frac{1 + 0 + M + E\,(M\,/\,4)}{7} = K + (0/7)$$

$$M + E\,(M\,/\,4) = 7\,(K - 1) + 7 + 0 - 1 = 7K' + 6\,K'' = K - 1$$

So $M + E\,(M\,/\,4)$ is congruent with 6 modulo 7.

Let's draw up the elements of the calendar 5 spectrum:

| 6 | 55 | 97 |
|---|----|----|
| 13 | 62 | 104 |

| 20 | 69 | 111 |
|----|----|-----|
| 27 | 76 | 118 |
| 34 | 83 | 125 |
| 41 | 90 | 132 |
| 48 |    |     |

$M + E (M / 4) = N$; $N$ belongs to the years of the above numbers. All the elements do not necessarily verify this equality, namely that $M + E (M / 4) = N$. Those which verify equality we call the nodes are effective by showing the years that renew the calendar 5.

If $N = 6$, then $M = 5$; $(5 + E (5/4)) = 5 + 1 = 6$.
If $N = 13$, then $M = 11$; $(11 + E (11/4)) = 11 + 2 = 13$.
If $N = 27$, then $M = 22$; $(22 + E (22/4)) = 22 + 5 = 27$.
If $N = 41$, then $M = 33$; $(33 + E (33/4)) = 33 + 8 = 41$.
If $N = 48$, then $M = 39$; $(39 + E (39/4)) = 39 + 9 = 48$.
If $N = 62$, then $M = 50$; $(50 + E (50/4)) = 50 + 12 = 62$.
If $N = 76$, then $M = 61$; $(61 + E (61/4)) = 61 + 15 = 76$.
If $N = 83$, then $M = 67$; $(67 + E (67/4)) = 67 + 16 = 83$.
If $N = 97$, then $M = 78$; $(78 + E (78/4)) = 78 + 19 = 97$.
If $N = 111$, then $M = 89$; $(89 + E (89/4)) = 89 + 22 = 111$.
If $N = 118$, then $M = 95$; $(95 + E (95/4)) = 95 + 23 = 118$.

The years 1905, 1911, 1922, 1933, 1939, 1950, 1961, 1967, 1978, 1989, and 1995 are represented by identical calendars from January 1 to December 31, the days of the week and the dates each correspond to each.

$$(5 - 11 - 22 - 33 - 39 - 50 - 61 - 67 - 78 - 89 - 95)$$

We see that:

$$11 - 5 = 6 \quad 39 - 33 = 6 \quad 67 - 61 = 6$$
$$22 - 11 = 11 \quad 50 - 39 = 11 \quad 78 - 67 = 11$$
$$33 - 22 = 11 \quad 61 - 50 = 11 \quad 89 - 78 = 11$$

We can continue and symmetrical elements appear. First of all:

$$33 - 5 = 28; \quad 61 - 33 = 28; \quad 89 - 61 = 28$$

The twenty-eight-year period appears, and then we notice that during this period, the reproduction takes place as follows: The first reproduction occurs six years after the first year, the second reproduction occurs eleven years after the first reproduction, and the third reproduction occurs eleven years after the second reproduction, therefore twenty-eight years after the first year of calendar 5. Calendar 5 is therefore characterized by the numbers 6 11 11 during a period of twenty-eight years. 6 11 11 are the pseudo-period or harmonics of this twenty-eight-year period. The ambulation of calendar 5 is therefore 6 11 11.

## Calendar 6

Let's go back to the empirical formula and find the day of the week that corresponds to January 1, 1906.

$$\frac{1 + 0 + 6 + E\,(6/4)}{7} = \frac{1 + 0 + 6 + 1 = (8/7)}{7} = 1 + (1/7)$$

So it's a Monday. Let's find out which years January 1 is a Monday.

$$\frac{1 + 0 + M + E\,(M/4)}{7} = K + (1/7)$$

$$M + E\,(M/4) = 7K + 1 - 1 = 7K + 0$$

$M + E\,(M/4)$ congruents with 0 modulo 7.

$M + E\,(M/4)$ belongs to the set of numbers, $N$, which is congruent with 0 modulo 7, but only a part of these numbers satisfies the equality of the formula $(M + E\,(M/4)) = N$. Let us draw up the list of numbers, $N$, which is congruent with 0 modulo 7; let's draw up the spectrum of the calendar 6.

The elements $N$ which satisfy $N = M + E\,(M\,/\,4)$ are the reproductions of calendar 6.

| 7  | 49 | 84  | 126 |
|----|----|-----|-----|
| 14 | 56 | 91  | 133 |
| 21 | 63 | 98  | 140 |
| 28 | 70 | 105 | 147 |
| 35 | 77 | 112 | 154 |
| 42 |    | 119 |     |

If $N = (M + E\,(M\,/\,4)) = 7$, then $M = 6$; $(6 + E\,(6/4)) = 6 + 1 = 7$.
If $N = 21$, then $M = 17$; $(17 + E\,(17/4)) = 17 + 4 = 21$.
If $N = 28$, then $M = 23$; $(23 + E\,(23/4)) = 23 + 5 = 28$.
If $N = 42$, then $M = 34$; $(34 + E\,(34/4)) = 34 + 8 = 42$.
If $N = 56$, then $M = 45$; $(45 + E\,(45/4)) = 45 + 11 = 56$.
If $N = 63$, then $M = 51$; $(51 + E\,(51/4)) = 51 + 12 = 63$.
If $N = 77$, then $M = 62$; $(62 + E\,(62/4)) = 62 + 15 = 77$.
If $N = 91$, then $M = 73$; $(73 + E\,(73/4)) = 73 + 18 = 91$.
If $N = 98$, then $M = 79$; $(79 + E\,(79/4)) = 79 + 19 = 98$.
If $N = 112$, then $M = 90$; $(90 + E\,(90/4)) = 90 + 22 = 112$.
If $N = 126$, then $M = 101$; $(101 + E\,(101/4)) = 101 + 25 = 126$.

The years 1906, 1917, 1923, 1934, 1945, 1951, 1962, 1973, 1979, 1990, and 2001 are represented by calendars absolutely identical to the 1906 calendar from January 1 to December 31, the days of the week, and the dates each correspond to each.

Let's try to see the structure of this set: 6-17-23-34-45-51-62-73-79-90-101.

We see that:

$$17 - 6 = 11 \quad 45 - 34 = 11 \quad 73 - 62 = 11$$
$$23 - 17 = 6 \quad 51 - 45 = 6 \quad 79 - 73 = 6$$
$$34 - 23 = 11 \quad 62 - 51 = 11 \quad 90 - 79 = 11$$

We can continue and symmetrical elements appear in this table. First of all:

$$34 - 6 = 28$$
$$62 - 34 = 28$$
$$90 - 62 = 28$$

The twenty-eight-year period appears, and then we notice that during this twenty-eight-year period, the reproduction takes place as follows: The first reproduction occurs eleven years after the first year of calendar 6, the second reproduction occurs six years after the first reproduction, and then the third reproduction occurs eleven years after the second reproduction, which is twenty-eight years after the first year of this calendar.

Calendar 6 is therefore characterized by the numbers 11–6–11 over twenty-eight years. 11–6–11 are the pseudo-periods or harmonics of the twenty-eight-year period; the ambulation of calendar 6 is therefore 11–6–11.

## Calendar 7

During the study of calendar 7, we saw that this calendar is identical to calendar 1.

## Calendar 8

Calendar 8 is a leap with two parts: the first part from January 1 to February 29 and the second part from March 1 to December 31.

For calendar 8 from January 1 to February 29, the formula must be applied with caution; in fact, the leap year is only effective from March 1.

From January 1 to February 29:

$$E\ (M\ /\ 4) =\ E\ ((8 - 1)\ /\ 4) =\ E\ (7/4) =\ 1$$

So using the formula, let's calculate the day of the week that corresponds to the date January 1, 1908.

$$\frac{1 + 0 + 8 + E\,(7/4)}{7} = \frac{1 + 0 + 8 + 1}{7} = 10/7 = 1 + (3/7)$$

So it's a Wednesday. The years in which January 1 is a Wednesday are the renewal of calendar 8. We are going to list all these years, or all the spectra of calendar 8.

$$\frac{1 + 0 + M + E\,((M - 1)\,/\,4)}{7} = K + (3/7)$$

$$1 + 0 + M + E\,((M - 1)\,/\,4) = 7K + 3$$

$$M + E\,((M - 1)\,/\,4) = 7K + 3 - 1 = 7K + 2$$

$M + E\,((M - 1)\,/\,4)$ is congruent with 2 modulo 7.

Let us draw up the set of the elements of the spectrum 8-1.

| 2 | 65 | 121 |
|---|---|---|
| 9 | 72 | 128 |
| 16 | 79 | 135 |
| 23 | 86 | 142 |
| 30 | 93 | 149 |
| 37 | 100 | 156 |
| 44 | 107 | 163 |
| 58 | 114 | |

If '$M + E\,((M - 1)\,/\,4)$' $= N = 4n$, then the first node of $M = 8$. Since $N = 4n$, $N$ is a multiple of 4, because $M$ must be leap.

If $M + E\,((M-1)\,/\,4) = N = 44$, then $M = 36$; $(36 + E\,(35/4)) = 36 + 8 = 44$.

If $N = 79$, then $M = 64$; $(64 + E\,(63/4)) = 64 + 15 = 79$.

If $N = 114$, then $M = 92$; $(92 + E\,(91/4)) = 92 + 22 = 114$.

The years 1908, 1936, 1964, and 1992 are all identical to the calendar 8-1. We notice that the spectrum of calendar 8, considering the part between January 1 and February 29, is absolutely identical to the spectrum of calendar 2.

Calendar 8 from March 1 to December 31

Let's find the day of the week that corresponds to March 1, 1908.

$$\frac{1 + 3 + 8 + E\,(8/4)}{7} = \frac{1 + 3 + 8 + 2}{7} = 14/7 = 2 + (0/7)$$

So it's a Sunday. We are going to establish the list of spectra of this calendar.

$M + E\,(M/4) = N$ is congruent with 3 modulo 7; we are going to establish the spectrum of calendar 8 from March 1 to December 31.

| 3 | 59 | 108 |
|---|---|---|
| 10 | 66 | 115 |
| 17 | 73 | 122 |
| 24 | 80 | 129 |
| 31 | 87 | 136 |
| 45 | 94 | 143 |
| 52 | 101 | 150 |

Nodal points:

If $(M + E\,(M/4) = N = 10$, so $M = 8$; $(8 + E\,(8/4)) = 8 + 2 = 10$.
If $N = 45$, then $M = 36$; $(36 + E\,(36/4)) = 36 + 9 = 45$.
If $N = 80$, then $M = 64$; $(64 + E\,(64/4)) = 64 + 16 = 80$.
If $N = 115$, then $M = 92$; $(92 + E\,(92/4) = 92 + 23 = 115$.

The years 1936, 1964, and 1992 are identical to 8, and the ambulation is twenty-eight years. First of all:

$$36 - 8 = 28;\ 64 - 36 = 28;\ \text{and}\ 92 - 64 = 28$$

The spectrum of calendar 8 from March 1 to December 31 is absolutely identical to calendar 3 from March 1 to December 31.

## Calendar 9

Let's apply the formula to find the day of the week that corresponds to the date of January 1, 1909.

$$\frac{1 + 0 + 9 + E\,(9/4)}{7} = \frac{1 + 0 + 9 + 2}{7} = 1 + (5/7)$$

So it's a Friday. The years of which January 1 is a Friday are the renewal of calendar 9. We are therefore going to establish the spectrum of these years.

$$\frac{1 + 0 + M + E\,(M/4)}{7} = K + (5/7)$$
$$1 + 0 + M + E\,(M/4) = 7K + 5$$
$$M + E\,(M/4) = 7K + 5 - 1 = 7K + 4$$

$M + E\,(M/4) = N$ is congruent with 4 modulo 7; let us draw up all the elements of the spectrum 9.

| | | |
|---|---|---|
| 4 | 53 | 102 |
| 11 | 60 | 109 |
| 18 | 67 | 116 |
| 25 | 74 | 123 |
| 32 | 81 | 130 |
| 39 | 88 | |
| 46 | 95 | |

We find that the spectrum of calendar 9 is similar to that of the first part of calendar 4, that is, from January 1 to February 28 of calendar 4.

Nodal points: $(M + E\,(M\,/\,4)) = N$:

If $(M + E\,(M\,/\,4) = N = 11$, so $M = 9$; $(9 + E\,(9/4)) = 9 + 2 = 11$.
If $N = 18$, so $M = 15$.
If $N = 35$, then $M = 26$.
If $N = 46$, then $M = 37$.
If $N = 53$, so $M = 43$.
If $N = 67$, then $M = 54$.
If $N = 81$, then $M = 65$.
If $N = 88$, then $M = 71$.
If $N = 102$, then $M = 82$.
If $N = 116$, then $M = 93$.

Years 9, 15, 26, 37, 43, 54, 65, 71, 82, and 93 are all identical to calendar 9.

The ambulation:

| 15 – 9 = 6 | 43 – 37 = 6 | 71 – 65 = 6 |
|---|---|---|
| 26 – 15 = 11 | 54 – 43 = 11 | 82 – 71 = 11 |
| 37 – 26 = 11 | 65 – 54 = 11 | 93 – 82 = 11 |

The calendar 9 ambulation is 6-11-11; at the same time, we have established that the periodicity of the calendar 9 is twenty-eight years.

## Calendar 10

Let's apply the formula to calculate the date of January 1, 1910:

$$\frac{1 + 0 + 10 + E\,(10/4)}{7} = \frac{1 + 0 + 10 + 2}{7} = 13/7 = 1 + (6/7)$$

So it's a Saturday. Years where January 1 is a Saturday are the calendar 10 renewal.

$$\frac{1 + 0 + M + E\,(M/4)}{7} = K + (6/7)$$

$$1 + 0 + M + E\,(M/4) = 7K + 6$$

$$M + E\,(M/4) = 7K + 6 - 1$$

$$M + E\,(M/4) = 7K + 5$$

$M + E\,(M/4) = N$ is congruent with 5 modulo 7. We will establish the elements of the calendar spectrum 10.

| 5 | 68 | 131 |
|---|---|---|
| 12 | 75 | 138 |
| 19 | 82 | 145 |
| 26 | 89 | 152 |
| 33 | 96 | 159 |
| 40 | 103 | 166 |
| 47 | 110 | 173 |
| 54 | 117 | 180 |
| 61 | 124 | |

Nodal points: $M + E\,(M/4) = N$:

If $N = 12$, so $M = 10$.
If $N = 26$, then $M = 21$.
If $N = 33$, then $M = 27$.
If $N = 47$, then $M = 38$.
If $N = 61$, then $M = 49$.
If $N = 68$, then $M = 55$.
If $N = 82$, then $M = 66$.
If $N = 96$, so $M = 77$.
If $N = 103$, so $M = 83$.
If $N = 117$, then $M = 94$; $(94 + E\,(94/4)) = 94 + 23 = 117$.

First of all, the periodicity is always twenty-eight years.

$$38 - 10 = 28;\ 66 - 38 = 28;\ \text{and}\ 94 - 66 = 28$$

Years 10, 21, 27, 38, 49, 55, 66, 77, 83, and 94 are all identical to calendar 10.

The ambulation:

$$21 - 10 = 11 \quad 49 - 38 = 11 \quad 77 - 66 = 11$$
$$27 - 21 = 6 \quad 55 - 49 = 6 \quad 83 - 77 = 6$$
$$38 - 27 = 11 \quad 66 - 55 = 11 \quad 94 - 83 = 11$$

The ambulation is as follows: 11-6-11.

## Important Note

We find that the spectrum of calendar 10 is exactly similar to that of calendar 4 from March 1 to December 31. The spectrum of calendar 4 is exactly similar to calendar 9 from January 1 to February 29.

## Calendar 11

This calendar is part of calendar 5, previously studied. They are identical from January 1 to December 31.

## Calendar 12

This leap calendar has two parts: the first part from January 1 to February 29 and the second part from March 1 to December 31. For calendar 12 from January 1 to February 29 here, the formula must be applied with care; indeed, the leap year is not effective until March 1. Let us calculate the day which corresponds to the date of January 1, 1912:

$$\frac{Q + I + M + E((M-1)/4)}{7} = K + (R/7$$

$$\frac{1 + 0 + 12 + E((12-1)/4) = 1 + 0 + 12 + E(11/4)}{7} = K2 + (1/7)$$

So it's a Monday; years for which January 1 is a Monday are all identical to calendar 12.

The spectrum of calendar 12 from January 1 to February 29:

$$\frac{1 + 0 + M + E\,((M-1)/4)}{7} = K + (1/7)$$

$$M + E\,((M-1)/4) = 7K + 1 - 1 = 7K$$

$M + E\,((M-1)/4) = N$ is congruent with 0 modulo 7. The spectrum of the years from January 1 to February 29 is reproduced by the following table:

| 7  | 42 | 81  |
|----|----|-----|
| 14 | 49 | 91  |
| 21 | 56 | 98  |
| 28 | 63 | 105 |
| 35 | 70 | 112 |
|    | 77 | 119 |

Nodal points: $M + E\,((M-1)/4) = N$; $M = 4n$.
If $N = 14$, then $M = 12$.
If $N = 49$, so $M = 40$.
If $N = 84$, then $M = 68$.
If $N = 119$, then $M = 96$.

$$(96 + E\,(95/4)) = 96 + 23 = 119$$

The calendars for the years 1912, 40, 68, and 96 are all identical. First, the periodicity:

$$40 - 12 = 28;\ 68 - 40 = 28;\ \text{and}\ 96 - 68 = 28$$

The periodicity is twenty-eight years. The ambulation is also twenty-eight years old:

$$40 - 12 = 28; 68 - 40 = 28; 96 - 68 = 28.$$

The spectrum of calendar 12 from January 1 to February 29 is similar to that of calendar 6.

For calendar 12 from March 1 to December 3, calculate the day which corresponds to the date of March 1, 1912.

$$\frac{1 + 3 + 12 + E(12/4)}{7} = \frac{1 + 3 + 12 + 3}{7} = 2 + (5/7)$$

So it's a Friday. We are going to establish the calendar 12 spectrum from March 1 to December 31.

$$\frac{1 + 3 + M + E(M/4)}{7} = K + (5/7)$$

$$4 + M + E(M/4) = 7K + 5; M + E(M/4) = 7K + 5 - 4 = 7K + 1$$

$M + E(M/4) = N$ is congruent with 1 modulo 7; we are going to establish the calendar 12 spectrum from March 1 to December 31.

| 1 | 43 | 85 |
|---|---|---|
| 8 | 50 | 92 |
| 15 | 57 | 99 |
| 22 | 64 | 106 |
| 29 | 71 | 113 |
| 36 | 78 | 120 |

Nodal points: $M + E(M/4) = N; M = 4n$

If $N = 15$, so $M = 12$.
If $N = 50$, so $M = 40$.
If $N = 85$, then $M = 68$.
If $N = 120$, so $M = 96$.

The years 1912, 40, 68, and 96 are all identical.
The periodicity and the ambulation are twenty-eight years.

## Calendars 13, 14, and 15

Calendar 13 is the identical renewal of calendar 2; calendar 14 is the identical renewal of calendar 3; and calendar 15 is the identical renewal of calendar 9.

## Calendar 16

Calendar 16 is a leap year. There are two parts to be considered carefully: the first part from January 1 to February 29. In the application of the formula, the leap year is only effective from March 1. Regarding the calendar 16 from January 1 to February 29, let's look for the day of the week that corresponds to the date of January 1, 1916.

$$\frac{Q + I + M + E\,((M-1)/4)}{7} = K + (R/7)$$

$$\frac{1 + 0 + 16 + E\,(15/4)}{7} = \frac{17 + 3 = 20/7}{7} = 2 + (6/7)$$

So it's a Saturday. Years where January 1 is a Saturday are all the same on calendar 16 from January 1 to February 28.

The spectrum of the 16-1 calendar:

$$\frac{1 + 0 + M + E\,((M-1)/4)}{7} = K + (6/7)$$

$$1 + 0 + M + E\,((M-1)/4) = 7K + 6; \quad M + E\,((M-1)/4) = 7K + 6 - 1$$
$$= 7K + 5$$

$M + E\,((M-1)/4) = N$ is congruent with 5 modulo 7; the spectrum of the calendar years 16-1 is 5, 12, 19, 26, 33, 40, 47, 54, 61, 68, 75, 82, 96, 103, 110, 117, and 124. We find that the spectrum of calendar 16-1 is identical to that of calendar 10 from January 1 to February 28. The nodes must be such that $M = 4n$.

If $N = 19$, then $M = 16$.
If $N = 54$, then $M = 44$.
If $N = 89$, then $M = 72$.
If $N = 124$, so $M = 100$.

The nodes where sixteen years exactly recurring from January 1 to February 29 are 16, 44, 72, and 100 (year 2000).

As for calendar 16, from March 1 to December 31, calculate the day of the week that corresponds to the date of March 1, 1916.

$$\frac{1 + 3 + 16 + E\,(16/4)}{7} = \frac{1 + 3 + 16 + 4}{7} = 3 + (3/7)$$

So it's a Wednesday. Let's look for the years for which March 1 is a Wednesday, that is to say, the years which exactly reproduce the years of the calendar 16 from March 1st to December 31.

$$\frac{1 + 3 + M + E\,(M/4)}{7} = K + (3/7)$$
$$M + E\,(M/4) = 7K + 3 - 1 - 3 = 7\,(K - 1) + 7 - 1 = 7K + 6$$

Therefore, $M + E\,(M/4) = N$ is congruent with 6 modulo 7. We will establish the spectrum of $M + E\,(M/4)$; we also know that $M$ is a leap year divisible by 4; therefore $M = 4n$, $n$ being a positive integer varying from 0 to infinity. Let us establish the spectrum of numbers which congruent with 6 modulo 7.

6, 13, 20, 27, 34, 41, 48, 55, 62, 69, 76, 83, 90, 97, 104, 111, 118, 125, 132, 139

The numbers representing the spectrum above are the different values that $N$ can take.

If $N = 20$, so $M = 16$.
If $N = 55$, then $M = 44$.
If $N = 90$, so $M = 72$.
If $N = 125$, so $M = 100$.

We notice that the above spectrum is exactly the same as the calendar 5 spectrum.

So calendars 16 and 5 are exactly the same from March 1 to December 31; years 16, 44, 72, and 100 (2000) are the years in which the calendar 16 reproduces similar to itself, so the nodes are 16, 44, 72, and 2000. The ambulation is 28, in fact:

$$44 - 16 = 28;\ 72 - 44 = 28;\ 100 - 72 = 28$$

The periodicity is also twenty-eight years.

## Calendars 17, 18, and 19

Calendar 17 is part of calendar 6; calendar 18 is part of calendar 1; and calendar 19 is part of calendar 2.

## Calendar 20

This calendar represents a leap year; the empirical formula should be applied with caution. There are two parts: the first from January 1 to February 29 and the second from March 1 to December 31. The first part of the calendar 20, 20-1, remains between January 1 and February 29. During this period, the number of leap years that are effective since 1901 is 4 (1904, 1908, 1912, 1916). Before March 1, year 20 is considered year 20-1; we calculate the day which corresponds to January 1, 1920.

$$\frac{Q + I + M + E\left((M-1)/4\right)}{7}$$
$$\frac{1 + 0 + 20 + E\left(19/4\right)}{7} = 25/7 = 3 + (4/7)$$

So it's a Thursday. Years where January 1 is a Thursday are identical to the 20-1 calendar. For the spectrum of the 20-1 calendar, let's look for the years in which January 1 is a Thursday; therefore, identical to calendar 20 from January 1 to February 28.

$$\frac{1 + 0 + M + E\,((M-1)\,/\,4)}{7} = K + (4/7)$$

$$M + E\,((M-1)\,/\,4) = 7K + 3$$

So these years are congruent with 3 modulo 7. Let us draw up the spectrum of these years:

| 3 | 52 | 101 |
|---|----|-----|
| 10 | 59 | 108 |
| 17 | 66 | 115 |
| 24 | 73 | 122 |
| 31 | 80 | 129 |
| 38 | 87 | 136 |
| 45 | 94 | 143 |

We know that $M$ is a leap, divisible by 4; $M = 4n$, where $n$ belongs to the set of numbers from 0 to infinity. With $M + E\,((M-1)\,/\,4) = N$, we will find the numbers representing the above spectrum and therefore the values of $N$.

If $N = 24$, then $M = 20$; $20 + E\,(19/4) = 20 + 4 = 24$.
If $N = 59$, then $M = 48$.
If $N = 94$, then $M = 76$.

So the nodes of the above spectrum are 20, 48, and 76. These years replicate calendar 20 from January 1 to February 29. Moreover, the spectrum of calendar 20 from January 1 to February 28 is similar to the spectrum of calendar 3.

As for the calendar 20 from March 1 to December 31, let's calculate the day of the week that corresponds to the date of March 1, 1920.

$$\frac{1 + 3 + 20 + E\,(20/4)}{7} = 29/7 = 4 + (1/7)$$

So it's a Monday; let's look for years in which March 1 is a Monday.

$$\frac{1 + 3 + M + E\,(M\,/\,4)}{7} = K + (1/7)$$
$$4 + M + E\,(M\,/\,4) = 7K + 1;\ M + E\,(M\,/\,4) = 7K + 1 - 4$$
$$M + E\,(M\,/\,4) = 7\,(K - 1) + 4$$

Therefore, $M + E\,(M\,/\,4) = N$ is congruent with 4 modulo 7; we call it the spectrum. Let's draw up the numbers that represent the spectrum, being made up of the following list.

$$4, 11, 18, 25, 32, 39, 46, 53, 60, 67, 74, 81, 88, 95, 102, 109, 116, \text{and } 123$$
$$M = 4n, \text{so } M + E\,(M\,/\,4)$$

Nodal points:

If $N = 25$, so $M = 20$.
If $N = 60$, then $M = 48$.
If $N = 95$, so $M = 76$.

We see that the nodal points which are the reproductions of the calendar are as follows: 20, 48, 76. We find that the periodicity is twenty-eight years.

Notes: The spectrum of calendar 20 is similar to that of calendar 9; the two calendars are identical from March 1 to December 31.

## Calendars 21, 22, and 23

Calendar 21 belongs to calendar 10; calendar 22 is a nodal point of calendar 5; and calendar 23 is a nodal point of calendar 6.

## Calendar 24

It is a leap. We will study it with care because it is made up of two parts. As for the first part, from January 1 to February 29, let's calculate the day of the week that corresponds to the date of January 1, 1924.

$$\frac{1 + 0 + 24 + E\left((24-1)/4\right)}{7} = \frac{1 + 24 + 5 = 30/7}{7} = 4 + (2/7)$$

So it's a Tuesday; let's look for the years when January 1 is a Tuesday.

$$\frac{1 + 0 + M + E\left((M-1)/4\right)}{7} = K + (2/7)$$

$$M + E\left((M-1)/4\right) = 7K + 2 - 1 = 7K + 1$$

$M + E(M/4)$ is congruent with 1 modulo 7; let us establish the spectrum which represents the number $N$:

| 1 | 43 | 85 |
|---|----|----|
| 8 | 50 | 92 |
| 15 | 57 | 99 |
| 22 | 64 | 106 |
| 29 | 71 | 113 |
| 36 | 78 | 120 |

With $M + E\left((M-1)/4\right) = N$:

If $N = 29$, then $M = 24$.
If $N = 64$, then $M = 52$.
If $N = 99$, then $M = 80$.

The nodes which are the reproduction of calendar 24 from January 1 to February 29 are 24, 52, 80. We note here that the spectrum of calendar 24 from January 1 to February 28 is similar to the spectrum of calendar 1.

As for the calendar 24, from March 1 to December 31, calculate the day of the week that corresponds to the date of March 1, 1924.

$$\frac{1 + 3 + 24 + E\,(24/4)}{7} = 34/7 = 4 + (6/7)$$

So it's a Saturday; let's look for years in which March 1ˢᵗ is a Saturday.

$$\frac{1 + 3 + M + E\,(M/4)}{7} = K + (6/7)$$

$$M + E\,(M/4) = 7K + 6 - 4 = 7K + 2$$

Therefore, $M + E\,(M/4) = N$ is congruent with 2 modulo 7; let us draw the spectrum of $M + E\,(M/4)$ which is congruent with 2 modulo 7, constituted by the following list:

| 2 | 44 | 86 |
|---|----|-----|
| 9 | 51 | 93 |
| 16 | 58 | 100 |
| 23 | 65 | 107 |
| 30 | 72 | 114 |
| 37 | 79 | |

With $M + E\,(M/4)$:

If $N = 30$, so $M = 24$.
If $N = 65$, then $M = 52$.
If $N = 100$, then $M = 80$.
So in the span of a century, years 24, 52, and 80 are reproductions of calendar 24 from March 1 to December 31.

$$52 - 24 = 28 \text{ and } 80 - 52 = 28$$

We note that the spectrum of calendar 24 from March 1 to December 31 is similar to that of calendar 2. So from March 1 to December 31, calendars 2 and 24 are absolutely identical.

## Calendars 25, 26, and 27

Calendar 25 is a nodal point of calendar 3; calendar 26 is a nodal point of calendar 9; and calendar 27 is a nodal point of calendar 10.

## Calendar 28

Calendar 28 is a leap year, consisting of two parts: the first from January 1 to February 29 and the second from March 1 to December 31. As for the first part, let's calculate the day of the week which corresponds to the date of January 1, 1928:

$$\frac{1 + 0 + 28 + E\,((28 - 1)}{7} = \frac{1 + 28 + 6 = 35/7}{7} = 5 + (0/7)$$

So it's a Sunday; let's look for the years for which January 1 is a Sunday, those which are the reproductions of year 28, the year 1928.

$$\frac{1 + 0 + M + E\,((M - 1)/4)}{7} = K + (0/7)$$

$$M + E\,((M - 1)/4) = 7K - 1 = 7(K - 1) = 7K + 6$$

These years are congruent with 6 modulo 7.

If $N$ is equal to the entity $M + E\,((M-1)/4)$, we can draw the spectrum that represents $N$ knowing that $N$ is a multiple of 4, so $M = 4n$.

| | | |
|---|---|---|
| 6 | 48 | 90 |
| 13 | 55 | 97 |
| 20 | 62 | 104 |
| 27 | 69 | 111 |
| 34 | 76 | 118 |
| 41 | 83 | 125 |

$N$, maybe, any number of this spectrum. With $M + E\,((M{-}1)\,/\,4) = N$:

If $N = 34$, then $M = 28$.
If $N = 69$, then $M = 56$.
If $N = 104$, then $M = 84$.

The nodes of the calendar 28 spectrum from January 1 to December 31 are 28, 56, and 84, in the space of a century; we find that the spectrum of calendar 28 from January 1 to February 28 is similar to calendar 5.

As for the second part of calendar 28, from March 1 to December 31, let's calculate the day of the week that corresponds to the date of March 1, 1928:

$$\frac{1 + 3 + 28 + E\,(28/4)}{7} = \frac{4 + 28 + 7 = 39/7}{7} = 5 + (4/7)$$

So it's a Thursday; let's look for all the years whose March 1 is a Thursday. Those years are the reproduction of the calendar 28 from March 1 to December 31:

$$\frac{1 + 3 + M + E\,(M\,/\,4)}{7} = K + (4/7)$$
$$M + E\,(M\,/\,4) = 7K + 4 - 4 = 7K + 0$$

Therefore, $M + E\,(M\,/\,4) = N$ is congruent with 0 modulo 7; draw the spectrum which corresponds to $M + E\,(M\,/\,4)$.

| 0 | 42 | 84 |
|---|---|---|
| 7 | 49 | 91 |
| 14 | 56 | 98 |
| 21 | 63 | 105 |
| 28 | 70 | 112 |
| 35 | 77 | 119 |

With $M + E\,(M\,/\,4) = N$:

If $N = 35$, then $M = 28$.
If $N = 70$, then $M = 56$.
If $N = 105$, then $M = 84$.

In the span of a century, nodes 28, 56, and 84 represent the reproduction of year 28 from March 1 to February 31. We note that the spectrum of calendar 28 from March 1 to December 31 is similar to Calendar 6; calendars 6 and 28 are identical from March 1 to December 31.

-     Calendar 29 is a nodal point in calendar 1.
-     Calendar 30 is a nodal point of calendar 2.
-     Calendar 31 is a nodal point of calendar 3.
-     Calendar 32 is a nodal point of calendar 4.
-     Calendar 33 is a nodal point of calendar 5.
-     Calendar 34 is a nodal point of calendar 6.
-     Calendar 35 is a nodal point of calendar 1.
-     Calendar 36 is a nodal point of calendar 8.
-     Calendar 37 is a nodal point of calendar 9.
-     Calendar 38 is a nodal point of calendar 10.
-     Calendar 39 is a nodal point of calendar 5.
-     Calendar 40 is a nodal point of calendar 12.

We find that all the calendars for the years after 1928 are reproductions of the calendars previously studied; this is normal because we had established that the periodicity is twenty-eight years.

For example, calendar 47 is the reproduction of calendar 2; calendar 55 is the reproduction of calendar 10; calendar 56 = calendar 28; calendar 73 = calendar 6; calendar 78 = calendar 5; calendar 80 = calendar 24; calendar 85 = calendar 1; calendar 86 = calendar 2; calendar 90 = calendar 6; calendar 95 = calendar 5; calendar 99 = calendar 9; and calendar 100 (2000) = calendar 16.

In summary, our study has shown that time is represented by 14 calendars which are the following:

- Non-leap calendars: 1, 2, 3, 5, 6, 9, 10
- Leap calendars: 4, 8, 12, 16, 20, 24, 28

The other years are just a century-long reproduction of the 14 calendars we previously looked at, for example, how to find the calendar that is reproduced by a given year:

1973 is the reproduction of which calendar?

$73 - 56 = 17$; 17 is one of the nodes of calendar 6, so 73 is a reproduction of calendar 6. The reference points are 28, 56, and 84. In the previous example, we can say that 73 comes after a leap year. The 1973 ambulation is therefore 6–11–11:

$$73 + 6 = 79; 79 + 11 = 90; 90 - 84 = 6$$

For example, what is the calendar which corresponds to the 1978 year?

78 comes two years after 76, which is a leap; therefore, the stroll of 78 is 11–6–11:

$$78 + 11 = 89 + 6 = 95, \text{and } 89 - 84 = 5$$

So this is calendar 5. Or we can think of it another way: $78 - 56 = 22$. 22 is a nodal point of calendar 5.

For example, which calendar does 1968 belong to?

$$68 - 56 = 12$$

1968 is the reproduction of the calendar 12.

# Explanation of How the "Time Diagram" Table Was Constructed

*Leap calendars*

March 1 of calendar 20 is a Monday, so we matched calendar 20 to Monday—the couple's leap year is listed in large numbers. From March 1 to December 31, calendar 20 is identical to calendar 9; from January 1 to February 28, calendar 20 is identical to calendar 3. Calendars 3 and 9 are not leap and their calendars are identified by small numbers. It is said that there is an interlacing between calendars 20 and 9 from March 1 to December 31 and that there is an interlacing between calendars 20 and 3 from January 1 to February 28.

March 1 of calendar 4 corresponds to a Tuesday; there is interlacing between calendars 4 and 9 from January 1 to February 28, and there is interlacing between calendars 4 and 10 from March 1 to December 31. Calendar 4 is represented by a large number, and calendars 9 and 10 are identified by small numbers.

We can now describe how the other calendars were placed about time—what has been said for calendars 20 and 4 is valid for the following calendars. Calendar 16 is placed on Wednesday; it interlaces with calendar 10 from January 1 to February 28, and calendar 16 interlaces with calendar 5 from March 1 to December 31. Calendar 28 is placed on Thursday; it interlaces with calendar 5 from January 1 to February 28, and calendar 28 interlaces with calendar 6 from March 1 to December 31. Calendar 12 is on Friday; it interlaces with calendar 6 from January 1 to February 28, and calendar 12 interlaces with calendar 1 from March 1 to December 31. Calendar 24 is on Saturday; it interlaces with calendar 1 from January 1 to February 28, and calendar 24 is interlaced with calendar 2 from March 1 to December 31.

Calendar 8 is on Sunday; it interlaces with calendar 2 from January 1 to February 28, and calendar 8 is interlaced with calendar 3 from March 1 to December 31.

This diagram is the graphic summary of all our studies from the beginning.

# Study of Index from Different Centuries

We have established that time is made up of fourteen entities that we have called calendars; the renewals of these fourteen calendars constitute the calendars that we know. Between the years 1901 and 2000, the month indices are zero for January, 3 for February, 3 for March, 6 for April, 1 for May, 4 for June, 6 for July, 2 for August, 5 for September, 0 for October, 3 for November, and 5 for December. For convenience, we present these indices as follows:

$$0, 3, 3, 6, 1, 4, 6, 2, 5, 0, 3, 5$$

We see that all these indices are between 0 and 6; for the other centuries, do the indices stay the same or not?

Let's calculate the day of the week that corresponds to the date of January 1, 1901.

$$\frac{1 + 0 + 1 + 0}{7} = 0 + (2/7)$$

So it's a Tuesday. If January 1, 1901, is a Tuesday, December 31, 1900, is a Monday. Write the research equation to determine the index for the month of December from 1801 to 1900.

$$\frac{31 + x + 100 + 24}{7} = x + 155 = K + (R / 7)$$

Indeed, there are one hundred years from 1801 to 1900, and there are twenty-four leap years between 1801 and 1900, $R$ being the row of the day of the week. Here, it's a Monday, so $R$ equals 1:

$$\frac{x + 155}{7} = x/7 + 22 + (1/7) = 22 + ((x + 1)/7) = 22 + (1/7)$$

We know that $K + R / 7$ is equal to $22 + 1/7$, so $x + 1$ is equal to 1; hence, $x$ is equal to 0. We found the index for the month of December from 1801 to 1900 equal to 0; we note that for a non-leap year, the day of the week of January 1 is the same as the day of the week of December 31. What he means is that January 1, 1900, is also a Monday. So we can calculate the index for the month of January 1900.

$$\frac{1 + x + 100 + 24}{7} = \frac{125 + x}{7} = 17 + ((x + 6) / 7)$$

Indeed, $(125/7) = ((119 + 6) / 7) = 17 + (6/7)$
We therefore have $((125 + x)/7) = 17 + (x + 6)/7$
We know that $((x + 6) / 7) = 1 + (1/7)$, so this gives us:

$$6 + x = 7 + 1 = 8; \; x = 8 - 6, \text{so } x = 2$$

The index for the month of January from 1801 to 1900 is equal to 2. Let us calculate the day of the week which corresponds to the date of January 1, 1801; we have just established that the index for the month of January of this century equals 2.

$$\frac{1 + 2 + 1 + 0}{7} = 0 + (4/7)$$

So it's a Thursday; since January 1, 1801, is a Thursday, therefore January 29, 1801, is also a Thursday. So February 1, 1801, is a Sunday. Let's calculate the index for the month of February.

$$\frac{1 + x + 1 + 0}{7} = K + (0/7)$$

$$\frac{1 + 2 + 1 + 0}{7} = ((x + 2)/7)$$

$((x + 2)/7) = K$ must be integer, and we know that $x$ is between 0 and 6; since $((x + 2)/7)$ is an integer, therefore $x$ equal to 5. The index of February from 1801 to 1900 equals 5.

Since February 1, 1801, is a Sunday, March 1 and March 29, 1801, are also a Sunday. Let us calculate the index for the month of March from 1801 to 1900.

$$\frac{1 + 2 + 1 + 0}{7} = K + (0/7)$$

$((x + 2)/7)$ must be an integer where $x$ equals 5; therefore, the index for the month of March from 1801 to 1900 equals 5. Since March 1, 1801, is a Sunday, March 29, 1801, is also a Sunday, so April 1, 1801, is a Wednesday. Let us calculate the index for the month of April from 1801 to 1900.

$$\frac{1 + x + 1 + 0}{7} = K + 3/7$$
$$((x + 2)/7) = K + 3/7$$

If $K = 0$, we have $((x + 2)/7) = 0 + (3/7)$.

$((x + 2)/7) = 3/7$ where $x = 1$. The index for the month of April from 1801 to 1900 equals 1; April 1 is a Wednesday, so May 1 is a Friday. Let us calculate the index for the month of May from 1801 to 1900.

$$\frac{1 + x + 1 + 0}{7} = K + 5/7$$

If $K = 0$, then $((x + 2) / 7) = 5/7$.

Hence $x = 3$; the index for the month of May from 1801 to 1900 equals 3. Since May 1, 1801, is a Friday, May 29, 1801, is also a Friday. June 1, 1801, is therefore a Monday; let's calculate the index for the month of June from 1801 to 1900.

$$\frac{1 + x + 1 + 0}{7} = K + 1/7$$

If $K = 1$, then $((x + 2) / 7) = 1 + 1/7 = 8/7$; $x + 2 = 8$

Hence $x = 6$; the index for the month of June from 1801 to 1900 equals 6. If June 1, 1801, is a Monday, June 29, 1801, is also a Monday; therefore, July 1, 1801, is a Wednesday. Let us calculate the index for the month of July from 1801 to 1900.

$$\frac{1 + x + 1 + 0}{7} = K + 3/7$$

This is the same equation as when looking for the index for the month of April from 1801 to 1900, so the index for the month of July from 1801 to 1900 equals 1. July 1, 1801, being a Wednesday, the $29^{\text{th}}$ of July 1801 is also a Wednesday; therefore, August 1 is a Saturday. Let us calculate the index for the month of August from 1801 to 1900.

$$\frac{1 + x + 1 + 0}{7} = K + 6/7$$

If $K = 0$, then $((x + 2) / 7) = 6/7$; $x + 2 = 6$

Hence $x = 4$, so the index for the month of August from 1801 to 1900 equals 4. August 1, 1801, being a Saturday, August 29, 1801, is also a Saturday; therefore, September 1, 1801, is a Tuesday.

$$\frac{1 + x + 1 + 0}{70} = K + 2/7$$

If $K = 0$, then $((x + 2) / 7) = 2/7$; $x + 2 = 2$

Hence $x = 0$, so the index for September from 1801 to 1900 equals 0.

September 1, 1801, being a Tuesday, September 29, 1801, is also a Tuesday.

Therefore, October 1, 1801, is a Thursday. Let us calculate the index for the month of October from 1801 to 1900.

$$\frac{1 + x + 1 + 0}{7} = K + 4/7$$

If $K = 0$, then $((x + 2) / 7) = 4/7$; $x + 2 = 4$

Hence $x = 2$, so the index for the month of October equals 2. October 1, 1801, being a Thursday, October 29, 1801, is also a Thursday; therefore, November 1, 1801, is a Sunday. Let us calculate the index for the month of November from 1801 to 1900.

$$\frac{1 + x + 1 + 0}{7} = K + 0/7$$

If $K = 1$, then $((x + 2) / 7) = 1 + 0/7 = 7/7$; $x + 2 = 7$

Hence $x = 5$, so the index for the month of November from 1801 to 1900 equals 5. November 1, 1801, being a Sunday, November 29, 1801, is also a Sunday; therefore, December 1, 1801, is a Tuesday. Let's calculate the index for the month of December from 1801 to 1900, verifying the answer shown at the beginning of this chapter.

$$\frac{1 + x + 1 + 0}{7} = K + 2/7$$

If $K = 0$, then $((x + 2) / 7) = 2/7$; $x + 2 = 0$

Hence $x = 0$, so the index for the month of December from 1801 to 1900 equals 0.

In summary, the indices for the months from 1801 to 1900 are as follows:

- January 2
- February 5
- March 5
- April 1
- May 3
- June 6
- July 1
- Aug 4
- September 0
- October 2
- November 5
- December 0

We present these indices as follows:

$$2, 5, 5, 1, 3, 6, 1, 4, 0, 2, 5, 0$$

Let us try to understand how these new indices, from 1801 to 1900, are deduced from the previous ones, that is to say, from the indices of the following century from 1901 to 2000. How can the index for January from 1801 to 1901, which equals 2, is deduced from the January index from 1901 to 2000, which equals 0?

So we go from 0 to 2, then we can write $0 + 2 = 2$. We see that by finding the index of a given month of a given century, we must decrease 2 from the index of the given month of the previous century.

Let's check that this rule is valid for all the indices. Thus, the indices for the month of February and March from 1801 to 1900 become for the months of February and March from 1801 to 1900:

- For the month of February from 1801 to 1900: $3 + 2 + = 5$.
- For the month of March from 1801 to 1900: $3 + 2 = 5$.

This is well verified by the indices of the 1$^{st}$ quarter from 1801 to 1900.

Let us examine the case of the second quarter of these centuries. The second quarter from 1901 to 2000 is for the months of April, May, and June. The indices are as follows: April = 6; May = 1; June = 4

If we apply the previous rule, the indices for the second quarter from 1801 to 1900 would be:

- April from 1801 to 1900: 6 + 2 = 8; 8 – 7 = 1
- May from 1801 to 1900: 1 + 2 = 3
- June from 1801 to 1900: 4 + 2 = 6

Here we find the indices that we have already calculated for the months of the second quarter from 1801 to 1900. Therefore, for the century from 1801 to 1900, the April index equals 1, the May index equals 3, and the June index equals 6. We notice here that the rule applies if we replace an index greater than 6 with its congruent modulo 7. We must bear in mind this precaution, namely, that all the indices are between 0 and 6.

As for the third quarter from 1801 to 1900, the indices are as follows: July = 1, August = 2, and September = 0.

For the third quarter from 1901 to 2000, the indices are the following: July = 6, August = 0, and September – 5.

We have understood that to go from indices of a lower century to indices of a later century, we must subtract 2 of the indices of the lower century. If the subtraction is not possible, from which we must subtract 2 is impossible, we replace the index with a number congruent with this modulo 7.

For example, the index for the month of July from 1801 to 1900 = 1, we cannot subtract 2 from 1. Therefore, we replace by its congruent modulo 7, that is to say, we replace 1 by 8, and after that, we can subtract 2 from 8, so we will have 6. 6 is the index for the month of July from 1901 to 2000. While the index for the month of July from 1801 to 1900 = 1. The index for the month of August from 1901 to 2000 equals 4 – 2 = 2. The index for the month of

September from 1901 to 2000 = 7 – 2 = 5; the zero of September from 1801 to 1900 is replaced by 7 which is congruent.

As for the fourth quarter from 1901 to 2000, the indices for this quarter of this century are deduced from the indices for the fourth quarter from 1801 to 1900. The indices for the fourth quarter from 1801 to 1900 are 2, 5, and 0 for October, November, and December, respectively. The indices for the fourth quarter from 1901 to 2000 are as follows:

- October = 2 – 2 = 0
- November = 5 – 2 = 3
- December = 7 – 2 = 5

We replaced 0 by 7, which is congruent of 0 modulo 7.

It was established on a previous page that January 1, 1801, is a Thursday, so December 31, 1800, is a Wednesday. Let's write the search equation:

$$\frac{\text{Date} + \text{month index} + \text{year} + \text{NLY}}{7} = K + R / 7$$

NLY is the number of leap years from 1701 to 1800, $K$ is an integer, and $R$ is the remainder of dividing the fraction by 7. As for December 31, 1800, we know that it is a Wednesday, if $x$ is the index of December from 1701 to 1800, we can write:

$$\frac{31 + x + 100 + 24}{7} = \frac{155 + x}{7} = 22 + ((1 + x) / 7)$$

We know it's a Wednesday, so $1 + x = 3$; $x = 2$.

Indeed, 1800 is not a leap year, which is why there are only twenty-four leap years from 1701 to 1800. We have just established that the index for the month of December from 1701 to 1800 equals 2. We remarked that the day of the week which corresponds to the date of January 1 of a non-leap year is the same day of the week which corresponds to the date of December 31 of the same year.

So the day of the week that corresponds to the date of January 1, 1800, is a Wednesday. So we can calculate the index for the month of January from 1701 to 1800. Let's use the search equation:

$$\frac{1 + x + 100 + 24}{7} = \frac{125 + x}{7} = \frac{119 + 6 + x}{7} = 17 + ((6 + x) / 7);$$

$$17 + ((6 + 1 - 1 + x) / 7) = 17 + ((1 + (x - 1) / 7) = 18 + (x - 1) / 7$$

Since we know that January 1, 1800, is a Wednesday, $x-1 = 3$, so $x = 3 + 1$; hence, $x = 4$; therefore, the index for the month of January from 1701 to 1800 equals 4.

Let's find the day of the week that corresponds to the date of January 1, 1701.

Back to the search equation:

$$\frac{1 + 4 + 1 + 0}{7} = ((0 + 6) / 7)$$

So it's a Saturday; January 29, 1701, is also a Saturday, so February 1, 1701, is a Tuesday. Let's calculate the index for the month of February from 1701 to 1800.

$$\frac{1 + x + 1 + 0}{7} - 0 + ((2 + x) / 7)$$

We know that February 1, 1701, is a Tuesday, so $2 + x = 2$; hence, $x = 0$. The index for February 1701 to 1800 equals 0. March 1, 1701, is a Tuesday, so March 29, 1701, is also a Tuesday; therefore, April 1, 1701, is a Friday. The index for March 1701 to 1800 equals the index for the month of February 1701 to 1800. Let us calculate the index for the month of April from 1701 to 1800.

$$\frac{1 + x + 1 + 0}{7} = 0 + ((2 + x) / 7)$$

We know that April 1, 1701, is a Friday, so $2 + x = 5$; hence, $x = 3$; the index for the month of April from 1701 to 1800 equals 3. With April 1, 1701, being a Friday, April 29, 1701, is also a Friday; therefore, May 1, 1701, is a Sunday. Let's go back to the search equation to find the index for the month of May from 1701 to 1800.

$$\frac{1 + x + 1 + 0}{7} = 0 + ((2 + x) / 7)$$

We know that May 1, 1701, is a Sunday, so $2 + x = 7$; hence, $x = 5$. The index for the month of May from 1701 to 1800 equals 5. May 1, 1701, being a Sunday, May 29, 1701, is also a Sunday; therefore, June 1, 1701, is a Wednesday. Let's calculate the index for the month of June from 1701 to 1800, using the search equation.

$$\frac{1 + x + 1 + 0}{7} = 0 + ((2 + x) / 7)$$

We know that June 1, 1701, is a Wednesday, so $x + 2 = 3$; hence, $x = 1$. The index for the month of June from 1701 to 1800 equals 1. June 1, 1701, being a Wednesday, June 29, 1701, is also a Wednesday; July 1, 1701, is therefore a Friday. Let us calculate the index for the month of July from 1701 to 1800.

$$\frac{1 + x + 1 + 0}{7} = 0 + ((2 + x) / 7)$$

We know that July 1, 1701, is a Friday, so $x + 2 = 5$; hence, $x = 3$; the index for the month of July 1701 to 1800 equals 3. Let us calculate the index for the month of August from 1701 to 1800. July 1, 1701, being a Friday, so July 29, 1701, is a Friday; therefore, August 1, 1701, is a Monday.

$$\frac{1 + x + 1 + 0}{7} = 0 + ((2 + x) / 7)$$

August 1, 1701, being a Monday, therefore:

$$0 + ((2 + x) / 7) = ((2 + 5 - 5 + x) / 7) = ((7 + (x - 5)) / 7)$$
$$= 1 + ((x - 5) / 7)$$

So $x$-5 = 1, hence $x$ = 6; therefore, the index for August from 1701 to 1800 equals 6.

Let us calculate the index for the month of September from 1701 to 1800. August 1, 1701, being a Monday, August 29, 1701, is also a Monday, so September 1, 1701, is a Thursday.

Let's go back to the research equation to determine the index for the month of September from 1701 to 1800.

$$\frac{1 + x + 1 + 0}{7} = 0 + ((2 + x) / 7)$$

We know that September 1, 1701, is a Thursday, so 2 + $x$ = 4; hence, $x$ = 2; the index for the month of September from 1701 to 1800 equals 2. Now calculate the index for the month of October from 1701 to 1800. September 1, 1701, is a Thursday, so September 29, 1701, is a Thursday; therefore, October 1, 1701, is a Saturday.

$$\frac{1 + x + 1 + 0}{7} = 0 + ((2 + x) / 7)$$

We know that October 1, 1701, is a Saturday.
So $x$ + 2 = 6, hence $x$ = 4. The index for October equals 4.

October 1, 1701, being a Saturday, so October 29, 1701, is also a Saturday; therefore, November 1, 1701, is a Tuesday. Let us calculate the index for the month of November from 1701 to 1800; we can write it using the search equation.

$$\frac{1 + x + 1 + 0}{7} = 0 + ((2 + x) / 7)$$

Since November 1, 1701, is a Tuesday, so $2 + x = 2$; hence, $x = 0$. The index for the month of November 1701 to 1800 equals 0. Now let's calculate the index for the month of December from 1701 to 1800. November 1, 1701, being a Tuesday, therefore November 29, 1701, is also a Tuesday; therefore, December 1, 1701, is a Thursday. Let's use the search equation to find the index for the month of December from 1701 to 1800:

$$\frac{1 + x + 1 + 0}{7} = 0 + ((2 + x) / 7)$$

We know that December 1, 1701, is a Thursday, so $2 + x = 4$; hence, $x = 2$; the index for the month of December from 1701 to 1800 equals 2. Let us summarize all the indices from 1701 to 1800: January equals 4, February = 0, March equals 0, April = 3, May equals 5, June = 1, July equals 3, August = 6, September equals 2, October = 4, November equals 0, and December = 2. We also write these indices as follows:

$$4, 0, 0, 3, 5, 1, 3, 6, 2, 4, 0, 2$$

We are going to calculate which day of the week corresponds to the date of January 1 of each leap calendar. Now we start with January 1 of calendar 4, being 1704:

$$\frac{1 + 4 + 4 + 0}{7} = 9/7 = 1 + 2/7$$

January 1, 1704, is a Tuesday; let's look for the day of the week that corresponds to January 1 of the calendar 8, 1708:

$$\frac{1 + 4 + 4 + 0}{7} = 14/7 = 2 + 0/7$$

January 1 of calendar 8 from 1701 to 1800 is Sunday. Now, let's see calendar 12, 1712:

$$\frac{1 + 4 + 12 + 2}{7} = 19/7 = 2 + 5/7$$

January 1, 1712, is a Friday. As for January 1 of calendar 16, 1716:

$$\frac{1 + 4 + 16 + 3}{7} + 3 = 24/7 = 3 + 3/7$$

January 1, 1716, is a Wednesday; as for January 1 of the calendar 20, 1720:

$$\frac{1 + 4 + 20 + 4}{7} = 29/7 = 4 + 1/7$$

January 1, 1720, is a Monday; as for January 1 of calendar 24:

$$\frac{1 + 4 + 24 + 5}{7} = 34/7 = 4 + 6/7$$

January 1, 1724, is a Saturday; as for January 1 of calendar 28, 1728:

$$\frac{11 + 4 + 28 + 6}{7} = 39/7 = 5 + 4/7$$

January 1, 1728, is a Thursday. We know that from January 1 to February 28, there is a day-to-day coincidence between leap calendars and non-leap calendars.

These coincidences are as follows:

| Leap calendars | Non-leap calendars |
| --- | --- |
| 4 | 9 |
| 8 | 2 |
| 12 | 6 |
| 16 | 10 |
| 20 | 3 |
| 24 | 1 |
| 28 | 5 |

Let's calculate the day of the week that corresponds to the date of March 1 of leap calendars, starting with March 1 of calendar 4, 1704:

$$\frac{1 + 0 + 4 + 1}{7} = 6/7 = 0 + 6/7$$

So it's a Saturday, as for March 1 of calendar 8, 1708:

$$\frac{1 + 0 + 8 + 2}{7} = 11/7 = 1 + 4/7$$

March 1, 1708, is a Thursday, as for March 1 of calendar 12:

$$\frac{1 + 0 + 12 + 3}{7} = 16/7 = 2 + 2/7$$

March 1, 1712, is a Tuesday, as for March 1 of calendar 16, 1716:

$$\frac{1 + 0 + 16 + 4}{7} = 21/7 = 3 + 0/7$$

March 1, 1716, is a Sunday, as for March 1 of the calendar 20, 1720:

$$\frac{1 + 0 + 20 + 5}{7} = 26/7 = 3 + 5/7$$

March 1, 1720, is a Friday, as for March 1 of calendar 24, 1724:

$$\frac{1 + 0 + 24 + 6}{7} = 31/7 = 4 + 3/7$$

March 1, 1724, is a Wednesday, as for March 1 of calendar 28, 1728:

$$\frac{1 + 0 + 28 + 7}{7} = 36/7 = 5 + 1/7$$

March 1, 1728, is a Monday.

We know that from March 1 to December 31, there is a day-to-day coincidence between leap calendars and non-leap calendars. These coincidences for the century from 1701 to 1800 are as follows:

| Leap calendars | Non-leap calendars |
|---|---|
| 4 | 10 |
| 8 | 3 |
| 12 | 1 |
| 16 | 5 |
| 20 | 9 |
| 24 | 2 |
| 28 | 6 |

**Table of the relationships between calendars and days of the week from January 1 to February 28 from 1701 to 1800**

|  | Leap calendars | Non-leap calendars |
|---|---|---|
| Monday | 20 | 3 |
| Tuesday | 4 | 9 |
| Wednesday | 16 | 10 |
| Thursday | 28 | 5 |
| Friday | 12 | 6 |
| Saturday | 24 | 1 |
| Sunday | 8 | 2 |

The left column indicates the days of the week that correspond to the date of January 1.

**Table of the relations between the calendars and the days of the week from March 1 to December 31 from 1701 to 1800**

|  | Leap calendars | Non-leap calendars |
|---|---|---|
| Monday | 28 | 6 |
| Tuesday | 12 | 1 |
| Wednesday | 24 | 2 |
| Thursday | 8 | 3 |
| Friday | 20 | 9 |
| Saturday | 4 | 10 |
| Sunday | 16 | 5 |

The left column indicates the days of the week that correspond to the date of March 1.

Let us calculate the indices for the years 1601 to 1700; we have understood the principle of calculating indices for each century, and we are going to continue this calculation for the other centuries. To avoid tedious repetitions of calculations for the reader, we invite the reader to admit the following indices for the years 1601 to 1700; he will also be able to verify the accuracy of these indices by applying the calculation principles that we have followed since the beginning of our study. These indices for the years 1601 to 1700 are as follows: 6 for the month of January, 2 for the month of February, 2 for the month of March, 5 for the month of April, 0 for the month of May, 3 for the month June, 5 for the month of July, 1 for the month of August, 4 for the month of September, 6 for the month of October, 2 for the month of November, and 4 for the month of December.

We can still write these indices as follows:

$$6, 2, 2, 5, 0, 3, 5, 1, 4, 6, 2, 4$$

We will calculate which day of the week corresponds to the date of January 1 of each leap calendar from 1601 to 1700; let's start with January 1, 1604:

$$\frac{1 + 6 + 4}{7} = \frac{11}{7} = 1 + \frac{4}{7}$$

January 1, 1604, is a Thursday, as for January 1 of calendar 8, 1608:

$$\frac{1 + 6 + 8 + 1}{7} = 16/7 = 2 + 2/7$$

January 1, 1608, is a Tuesday, as for January 1 of calendar 12, 1612:

$$\frac{1 + 6 + 12 + 2 = 21}{7} = 3 + 0/7$$

January 1, 1612, is a Sunday, as for January 1 of the calendar 16, 1616:

$$\frac{1 + 6 + 16 + 3}{7} = 26/7 = 3 + 5/7$$

January 1, 1616, is a Friday. As for January 1 of calendar 20, 1620:

$$\frac{1 + 6 + 20 + 4}{7} = 31/7 = 4 + 3/7$$

January 1, 1620, is a Wednesday, as for January 1 of the calendar 24, 1624:

$$\frac{1 + 6 + 24 + 5}{7} = 36/7 = 5 + 1/7$$

January 1, 1624, is a Monday, as for January 1 of calendar 28, 1628:

$$\frac{1 + 6 + 28 + 6}{7} = 41/7 = 5 + 6/7$$

January 1, 1628, is a Saturday. We can summarize these results in the following table, taking into account the interlocking between leap calendars and non-leap calendars; we know that these hugs are from January 1 to February 28.

### January 1 to February 28 1601 to 1700

|  | Leap calendars | Non-leap calendars |
|---|---|---|
| Monday | 24 | 1 |
| Tuesday | 8 | 2 |
| Wednesday | 20 | 3 |
| Thursday | 4 | 9 |
| Friday | 16 | 10 |
| Saturday | 28 | 5 |
| Sunday | 12 | 6 |

The days of the week in the above table correspond to the date of January 1 from 1601 to 1700. Let us calculate the days of the week that correspond to the date of March 1 for the century from 1601 to 1700.

Let's start with March 1 of calendar 4, 1604:

$$\frac{1 + 2 + 4 + 1}{7} = 8/7 = 1 + 1/7$$

March 1, 1604, is a Monday, as for March 1 of calendar 8, 1608:

$$\frac{1 + 2 + 8 + 2}{7} = 13/7 = 1 + 6/7$$

March 1, 1608, is a Saturday, as for March 1 of calendar 12, 1612:

$$\frac{1 + 2 + 12 + 3}{7} = 18/7 = 2 + 4/7$$

March 1, 1612, is a Thursday, as for March 1 of calendar 16, 1616:

$$\frac{1 + 2 + 16 + 4}{7} = 23/7 = 3 + 2/7$$

March 1, 1616, is a Tuesday, as to the calendar 20, 1620:

$$\frac{1 + 2 + 20 + 5}{7} \quad 28/7 = 4 + 0/7$$

March 1, 1620, is a Sunday, as to the calendar 24, 1624:

$$\frac{1 + 2 + 24 + 6}{7} \quad 33/7 = 4 + 5/7$$

March 1, 1624, is a Friday, as to calendar 28, 1628:

$$\frac{1 + 2 + 28 + 7}{7} \; 38/7 = 5 + 3/7$$

March 1, 1628, is a Wednesday. We know that from March 1 to December 31, there are the following coincidences between leap calendars and non-leap calendars. Calendar 4 and calendar 10 correspond from March 1 to December 31.

$$8 = 3, 12 = 1, 16 = 5, 20 = 9, 24 = 2, 28 = 6$$

We can summarize these results in the following table:

**March 1 to December 31, 1601, to 1700**

|  | Leap calendars | Non-leap calendars |
|---|---|---|
| Monday | 4 | 10 |
| Tuesday | 16 | 5 |
| Wednesday | 28 | 6 |
| Thursday | 12 | 1 |
| Friday | 24 | 2 |
| Saturday | 8 | 3 |
| Sunday | 20 | 9 |

Using this last table, we can plot the time from 1601 to 1700.

As we did to calculate the dates that correspond to the calendars from 1601 to 1700, we will also avoid the tedious calculations here. But we take into account that the year 1600 is a leap year.

In the previous pages, we had calculated that January 1, 1624, is a Monday, and Monday also corresponds to the date of January 1 by a coincidence of calendars 24 and 1 from January 1 to February 28. So December 31, 1600, is a Sunday. Let us calculate the index for the month of December from 1501 to 1600.

$$(31 + x + 100 + 25)/7 = ((156 + x)/7 = 22 + ((x + 2)/7)$$
$$= 22 + ((x - 5 + 5 + 2)/7) = 22 + ((x + 5)/7)$$
$$= 23 + ((x - 5)/7)$$

We know that December 31, 1600, is a Sunday, so $x - 5 = 0$; hence, $x = 5$; therefore, the index for the month of December from 1501 to 1600 equals 5. January 1, 1600, is a Sunday, since the year 1600 is not a leap year; let's calculate the index for the month of January from 1501 to 1600.

$$\frac{1 + x + 100 + 24}{7} \quad ((125 + x)/7 = 17 + ((x + 6)/7)$$

We know that January 1, 1600, is a Sunday, so $x + 6 = 6$; hence, $x = 0$. The index for the month of January from 1501 to 1600 equals 0.

Let us calculate the day which corresponds to the date of January 1, 1501:

$$\frac{1 + 0 + 1 + 0}{7} = 0 + 2/7$$

It's a Tuesday. January 29, 1501, is also a Tuesday; therefore, February 1 is a Friday. Let us calculate the index for the month of February from 1501 to 1600.

$$\frac{1 + x + 1 + 0}{7} = (2 + x)/7$$

We know that February 1 is a Friday, so $2 + x = 5$; hence, $x = 3$. The February index from 1501 to 1600 equals 3. The March index from 1501 to 1600 equals also 3.

March 1, 1501, being a Friday, March 29, 1501, is also a Friday; therefore, April 1, 1501, is a Monday. Let's calculate the index for the month of April from 1501 to 1600.

$$\frac{1 + x + 1 + 0}{7} (2 + x) / 7 = 1 + (x - 5) / 7$$

We know that April 1, 1501, is a Monday, so $x - 5 = 1$; hence, $x = 6$. The index for April from 1501 to 1600 equals 6. April 29, 1501, is also a Monday, so May 1, 1501, is a Wednesday; let's calculate the index for the month of May from 1501 to 1600:

$$\frac{1 + x + 1 + 0}{7} (2 + x) / 7 = 0 + (x + 2) / 7$$

We know that May 1, 1501, is a Wednesday, so $x + 2 = 3$; hence, $x = 1$; the index for the month of May from 1501 to 1600 equals 1. With May 1, 1501, being a Wednesday, May 29, 1501, is also a Wednesday; therefore, June 1, 1501, is a Saturday. Let us calculate the index for the month of June from 1501 to 1600.

$$\frac{1 + x + 1 + 0}{7} = \frac{(2 + x)}{7}$$

We know that June 1, 1501, is a Saturday, so $2 + x = 6$; hence, $x = 4$; the index for the month of June from 1501 to 1600 equals 4. With June 1, 1501, being a Saturday, June 29, 1501, is also a Saturday; July 1, 1501, is therefore a Monday. Let us calculate the index for the month of July from 1501 to 1600.

$$\frac{1 + x + 1 + 0}{7} (2 + x) / 7 = 1 + (x - 5) / 7$$

We know that July 1, 1501, is a Monday, so $x - 5 = 1$; hence, $x = 6$. The index for the month of July from 1501 to 1600 equals 6. July 1, 1501, being a Monday, so July 29, 1501, is also a Monday; August 1, 1501, is therefore a Thursday. Let's calculate the index for August from 1501 to 1600.

$$\frac{1 + x + 1 + 0}{7} = (2 + x) / 7$$

We know that August 1 is a Thursday, so $2 + x = 4$; hence, $x = 2$. The index for August from 1501 to 1600 equals 2. August 1 being a Thursday, August 29 is also a Thursday, so September 1 is a Sunday.

Let us calculate the index for the month of September from 1501 to 1600.

$$\frac{1 + x + 1 + 0}{7} = (2 + x) / 71 + (x - 5) / 7$$

We know that September 1, 1501, is a Sunday, so $x - 5 = 0$; hence, $x = 5$. The index for September equals 5. September 1, 1501, being a Sunday, September 29, 1501, is also a Sunday; therefore, October 1 is a Tuesday. We note that the date of October 1 arrives on the same day of the week as January 1; therefore, the index for October is the same as the index for January. The index for the month of October from 1501 to 1600 therefore equals 0. Since October 1, 1501, is a Tuesday, October 29, 1501, is also a Tuesday, so November 1, 1501, is a Friday. But we note here that November 1 arrives on the same day of the week as February 1 and March 1; therefore, the index for the month of November is the same as the indexes for the months of February and March. The index for the month of November from 1501 to 1600 equals 3. November 1, 1501, being a Friday, so November 29, 1501, is also a Friday; therefore, December 1, 1501, is a Sunday. The date of December 1 occurs on the same day of the week as September 1, so the index for December is the same as the index for September. The index for the month of December from 1501 to 1600 equals 5.

We can summarize all the indices for the years 1501 to 1600: January index 0, February index 3, March index 3, April index 6, May index 1, June index 4, July index 6, August index 2, September index 5, October index 0, November index 3, and December index 5. These indices can be presented differently:

$$0, 3, 3, 6, 1, 4, 6, 2, 5, 0, 3, 5$$

We note that these indices are exactly the same as the indices for the months from 1901 to 2000.

In the previous pages, we have established that January 1, 1501, is a Tuesday; therefore, December 31, 1500, is a Monday. Thus, we can calculate the index for the month of December from 1401 to 1500.

$$\frac{31 + x + 100 + 24}{7} = 31 + x + 100 + 24$$

We know that December 31, 1500, is a Monday, so $x + 1 = 1$; hence, $x = 0$. The index for the month of December from 1401 to 1500 equals 0.

The year 1500, not being a leap, January 1, 1500, is also a Monday, as the day of the week that corresponds to December 31, 1500.

Let us calculate the index for the month of January 1401 to 1500.

January 1, 1500, being a Monday.

$$\frac{1 + x + 100 + 24}{7} = (125 + x)/7 = 17 + (x + 6)/7 = \frac{x - 1 + 1 + 6 + 17}{7}$$
$$= \frac{17 + (x - 1) + 7}{7} = 18 + (x - 1)/7$$

$x - 1 = 1$, so $x = 2$. The index for the month of January 1401 to 1500 equals 2.

Let's calculate the day of the week that corresponds to the date of January 1, 1401.

$$(1 + 2 + 1 + 0)/7 = 4/7$$

Since January 1, 1401, is a Thursday, January 29, 1401, is also a Thursday. February 1 is therefore a Sunday; let's calculate the index for the month of February 1401 to 1500.

$$\frac{1 + x + 1 + 0}{7} = \frac{1 + x + 1 + 0}{7} = \frac{x - 5 + 5 + 2}{7} = 1 + (x - 5)/7$$

$x - 5 = 0$; hence, $x = 5$; the index for the month of February 1401 to 1500 equals 5. We know that the index for the month of March is the same as that of February, so the index for the month of March from 1401 to 1500 equals 5. March 1, 1401, is also a Sunday, so March 29, 1401, is a Sunday. April 1, 1401, is a Wednesday; let's calculate the index for the month of April from 1401 to 1500.

$$\frac{1 + x + 1 + 0}{7} = (x + 2)/7$$

Since April 1, 1401, is a Wednesday, $x + 2 = 3$; hence, $x = 1$; the index for the month of April equals 1.

April 29, 1401, is also a Wednesday, so May 1, 1401, is a Friday; let's calculate the index for the month of May from 1401 to 1500.

$$\frac{1 + x + 1 + 0}{7} = (x + 2)/7$$

Since May 1, 1401, is a Friday, $x + 2 = 5$; hence, $x = 3$; the index for the month of May from 1401 to 1500 equals 3.

May 29, 1401, is a Friday, so June 1, 1401, is a Monday. Let's calculate the index for the month of June from 1401 to 1500.

$$\frac{1 + x + 1 + 0}{7} = \frac{x + 2}{7} = \frac{x - 5 + 5 + 2}{7} = 1 + (x - 5)/7$$

Since June 1, 1401, is a Monday, $x - 5 = 1$; hence, $x = 6$. The index for the month of June 1401 to 1500 equals 6. June 29, 1401, is also a Monday, so July 1, 1401, is a Wednesday; let's calculate the index for the month of July from 1401 to 1500.

$$\frac{1 + x + 1 + 0}{7} = (x + 2) / 7$$

Since July 1, 1401, is a Wednesday, $x + 2 = 3$; hence, $x = 1$; the index for the month of July from 1401 to 1500 equals 1. July 29, 1401, is also a Wednesday, so August 1, 1401, is a Saturday; let's calculate the index for August from 1401 to 1500.

$$\frac{1 + x + 1 + 0}{7} = (x + 2) / 7$$

Since August 1, 1401, is a Saturday, $x + 2 = 6$; hence, $x = 4$; the index for August from 1401 to 1500 equals 4. August 29, 1401, is also a Saturday, so September 1, 1401, is a Tuesday; let's calculate the index for the month of September from 1401 to 1500.

$$\frac{1 + x + 1 + 0}{7} = (x + 2) / 7$$

Since September 1, 1401, is a Tuesday, $x + 2 = 2$; hence, $x = 0$; the index for the month of September from 1401 to 1500 equals 0. September 29, 1401, is also a Tuesday, so October 1, 1401, is a Thursday; let's calculate the index for the month of October from 1401 to 1500.

$$\frac{1 + x + 1 + 0}{7} = (x + 2) / 7$$

Since October 1, 1401, is a Thursday, $x + 2 = 4$; hence, $x = 2$; the index for the month of October from 1401 to 1500 equals 2.

October 29, 1401, is also a Thursday, so November 1, 1401, is a Sunday; let's calculate the index for the month of November from 1401 to 1500.

$$\frac{1 + x + 1 + 0}{7} = \frac{x + 2}{7} = \frac{x - 5 + 5 + 2}{7} = 1 + (x - 5)/7$$

Since November 1, 1401, is a Sunday, $x - 5 = 0$; hence, $x = 5$; the index for the month of November from 1401 to 1500 equals 5. November 29 is also a Sunday, so December 1, 1401 is a Tuesday; let's calculate the index for the month of December from 1401 to 1500.

$$\frac{1 + x + 1 + 0}{7} = (x + 2)/7$$

Since December 1, 1401, is a Tuesday, $x + 2 = 2$; hence, $x = 0$; the index for the month of December from 1401 to 1500 equals 0. In summary, the indexes for the months from 1401 to 1500 are as follows: January = 2, February = 5, March = 5, April = 1, May = 3, June = 6, July = 1, August = 4, September = 0, October = 2, November = 5, and December = 0. We can write the indices in a more convenient way for the view by grouping them three by three:

$$2, 5, 5, 1, 3, 6, 1, 4, 0, \ 2, 5, 0$$

Thus, we see that these are exactly the same indices as those for the years 1801 to 1900.

We have made in the previous pages the same remarks concerning the similarity between the indices of the years from 1501 to 1600 on the one hand and the indices of the years from 1901 to 2000. We realize that if we continue to calculate the indices for the centuries below 1401 with the same methods of calculation, we will simply have the repetition of the results that we have already obtained. We thus see appearing a periodicity of four hundred years for the indices of the centuries. To confirm the results stated above, let us calculate the indices for the years 2001 to 2100. December 31, 2000 is:

$$\frac{31 + 5 + 100 + 25}{7} = 161/7 = 23 + 0/7$$

December 31, 2000, is a Sunday, so January 1, 2001, is a Monday; let's calculate the index for the month of January from 2001 to 2100.

$$\frac{1 + x + 1 + 0}{7} = \frac{x + 2}{7} = \frac{x - 5 + 5 + 2}{7} = 1 + (x - 5)/7$$

Since January 1, 2001, is a Monday, $x - 5 = 1$; hence, $x = 6$; the index for the month of January 2001 equals 6.

January 29, 2001, is also a Monday, so February 1, 2001, is a Thursday; let's calculate the index for the month of February from 2001 to 2100.

$$\frac{1 + x + 1 + 0}{7} = (x + 2)/7$$

Since February 1, 2001, is a Thursday, $x + 2 = 4$; hence, $x = 2$; the index for February from 2001 to 2100 equals 2. March 1, 2001, is also a Thursday, and we know that the index for March is the same as the index for February, so the index from March 2001 to 2100 equals 2.

March 29, 2001, is also a Thursday, so April 1, 2001, is a Sunday; let's calculate the index for the month of April from 2001 to 2100.

$$\frac{1 + x + 1 + 0}{7} = \frac{x + 2}{7} = \frac{x - 5 + 5 + 2}{7} = 1 + (x - 5)/7$$

Since April 1, 2001, is a Sunday, $x - 5 = 0$; hence, $x = 5$; the index for the month of April from 2001 to 2100 equals 5. April 29, 2001, is also a Sunday, so May 1, 2001, is a Tuesday; let's calculate the index for the month of May from 2001 to 2100.

$$\frac{1 + x + 1 + 0}{7} = (x + 2)/7$$

Since May 1, 2001, is a Tuesday, $x + 2 = 2$; hence, $x = 0$; the index for the month of May from 2001 to 2100 equals 0. May 29, 2001, is also a Tuesday, so June 1, 2001, is a Friday; let's calculate the index for the month of June from 2001 to 2100.

$$\frac{1 + x + 1 + 0}{7} = (x + 2) / 7$$

Since June 1, 2001, is a Friday, $x + 2 = 5$; hence, $x = 3$; the index for the month of June from 2001 to 2100 equals 3. June 29, 2001, is also a Friday, so July 1, 2001, is a Sunday; let's calculate the index for the month of July from 2001 to 2100.

$$\frac{1 + x + 1 + 0}{7} = \frac{x + 2}{7} = \frac{x - 5 + 5 + 2}{7} = 1 + (x - 5) / 7$$

Since July 1, 2001, is a Sunday, $x - 5 = 0$; hence, $x = 5$; the index for the month of July from 2001 to 2100 equals 5. July 29, 2001, is also a Sunday, so August 1, 2001, is a Wednesday; let's calculate the index for the month of August from 2001 to 2100.

$$\frac{1 + x + 1 + 0}{7} = (x + 2) / 7$$

Since August 1, 2001, is a Wednesday, $x + 2 = 3$; hence, $x = 1$; the index for August from 2001 to 2100 equals 1.

August 29, 2001, is also a Wednesday, so September 1, 2001, is a Saturday; let's calculate the index for the month of September from 2001 to 2100.

$$\frac{1 + x + 1 + 0}{7} = (x + 2) / 7$$

Since September 1, 2001, is a Saturday, $x + 2 = 6$; hence, $x = 4$; the index for the month of September from 2001 to 2100 equals 4. September 29, 2001, is also a Saturday, so October 1, 2001, is a

Monday; let's calculate the index for the month of October from 2001 to 2100.

$$\frac{1 + x + 1 + 0}{7} = \frac{x + 2}{7} = \frac{x - 5 + 5 + 2}{7} = 1 + (x - 5)/7$$

Since October 1, 2001, is a Monday, $x - 5 = 1$; hence, $x = 6$; the index for the month of October from 2001 to 2100 equals 6. October 29, 2001, is also a Monday, so November 1, 2001, is a Thursday; let's calculate the index for the month of November from 2001 to 2100.

$$\frac{1 + x + 1 + 0}{7} = (x + 2)/7$$

Since November 1, 2001, is a Thursday, $x + 2 = 4$; hence, $x = 2$; the index for the month of November from 2001 to 2100 equals 2. November 29, 2001, is also a Thursday, so December 1, 2001, is a Saturday; calculate the index for the month of December from 2001 to 2100.

$$\frac{1 + x + 1 + 0}{7} = (x + 2)/7$$

Since December 1, 2001, is a Saturday, $x + 2 = 6$; hence, $x = 4$; the index for the month of December from 2001 to 2100 equals 4. Thus, if we summarize the indexes for the month from 2001 to 2100, they are as follows: January = 6, February = 2, March = 2, April = 5, May = 0, June = 3, July = 5, August = 1, September = 4, October = 6, November = 2, and December = 4. Let's write another way by grouping the months three by three for convenience:

6, 2, 2, 5, 0, 3, 5, 1, 4, 6, 2, 4

Thus, we have just verified that the indices for centuries 1601 to 1700 on the one hand, and 2001 to 2100 on the other hand, are absolutely identical; there is therefore a periodicity of four hundred years for the month indices. Let's recap all the month indices for the seven centuries.

| 1401–1500 | 255 | 136 | 140 | 250 |
|---|---|---|---|---|
| 1501–1600 | 033 | 614 | 625 | 035 |
| 1601–1700 | 622 | 503 | 514 | 624 |
| 1701–1800 | 400 | 351 | 362 | 402 |
| 1801–1900 | 255 | 136 | 140 | 250 |
| 1901–2000 | 033 | 614 | 625 | 035 |
| 2001–2100 | 622 | 503 | 514 | 624 |

The four-hundred-year periodicity appears more clearly here. We notice that the series for each century is deduced from the previous one according to an immutable rule, taking certain considerations into account of course; whether the end of the preceding century is a leap or not, and by replacing certain indices by their congruent modules 7, we can find them. Since there are no negative clues.

For example, how to go from indices of centuries 1401 to 1500 to indices of centuries 1501 to 1600?

Take for example the first three indices of the first three months from 1401 to 1500. We have January with index 2, February with 5, and March with 5.

When we go to the century 1501 to 1600, we see January for an index 0, February with 3, and March with 3; we can clearly see that the January from 1501 to 1600 is deduced from 1401 to 1500 by subtracting the index for the month of January (1401 to 1500) minus two makes the index for the month of January (1501 to 1600).

$$2 - 2 = 0$$

The index for March from 1401 to 1500 minus 2 makes the index for March from 1501 to 1600. If we take the index for April from 1401 to 1500 = 1, here arithmetic subtraction can not happen automatically. We replace 1 by 8 which is congruent modulo 7. The index of April 1401 to 1500 minus two makes the index of April 1501 to 1600.

$$8 - 2 = 6$$

The rule is that we subtract 2 from the index of the previous century to obtain the index of the following century.

We note that 1500 is not a leap century. If the pivotal century, a century from which we are emerging, is a leap, we subtract the number 1 from the index of the month of the past centuries to obtain the index of the corresponding month of the following centuries.

Consider the first three months of the century 1501 to 1600 and the first three months of the century 1601 to 1700. For 1501 to 1600, January = 0, February = 3; and March = 3. The index for January 1601 to 1700 will be the index for January 1501 minus 1, 0 which is 7 by its congruent modulo 7, seen as 7 – 1 = 6; therefore, the index for January 1601 to 1700 equals 6.

We are going to unwind the tangles of time. We will try to see what use can be made of the results obtained. Like the different radiations that make up white light, time is made up of several entities that we have called calendars.

We have spent a long time demonstrating the existence of these calendars within a given century. Next, we proceeded to establish evidence of a repeating cycle of four centuries.

The days of the week have some beneficial effects and some evil ones. In the pure language of the Fulani, Monday is *Amnde*, which means "brings good luck," and Tuesday is *Moobaare*, which means disaster. We know of families who for nothing in the world will take a trip on Friday. All of these popular beliefs cannot be taken into account because they are not based on any reliable statistics made by credible people on representative samples. Indeed, we know poor people who are well-born on Saturdays. We still know of perfect morons who were born on Tuesday or Friday. We know that we have made trips on Tuesday or Friday several times and nothing particularly bad has happened to us. Also, we say for a fact to have any chance of becoming law, that fact must be repeated with a frequency of at least 70 percent under absolutely identical conditions.

For example, are years starting with a Saturday good or not?

In this case, the years in which January 1 is a Saturday is as follows:

- For the period 1901 to 2000, these are calendar years 10 or calendar years 16.
- From 1801 to 1900 years for which January 1 is a Saturday are the years belonging to calendar 3 or calendar 20. For calendar 3: 1803, 1814, 1825, 1831, 1842, 1853, 1859, 1870, 1881, 1887, and 1898, or eleven in number.
- The years which belong to the calendar 20 are as follows: 1820, 1848, and 1876, making a total of three. That's a total of fourteen from 1801 to 1900.
- From 1901 to 2000, the years belonging to calendar 10 are the following: 1910, 1921, 1927, 1938, 1949, 1955, 1966, 1977, 1983, and 1994, making a total of ten.
- From 1901 to 2000, the years that belong to calendar 16 are as follows: 1916, 1944, 1972, and 2000, making a total of four. This is a total of fourteen from 1901 to 2000.

In the twentieth century, we have fourteen calendars with January 1 being a Saturday, and in the nineteenth century, we also have fourteen calendars with January 1 being a Saturday. So we have a total of twenty-eight calendars for the two centuries—the nineteenth and the twentieth—and with the records preserved by history for all the hearings of the globe, we can verify that a year beginning on a Saturday gives that year the unenviable character of a dreadful year of guns and blood. Of the twenty-eight years, at least twenty have been verified as catastrophic years; 70 percent at least can no longer be dismissed out of hand by this popular belief or by this claim of the mystics. We can now ask this: Catastrophic, exactly what does that mean? And are there things that happen every year that can be discerned as catastrophic? This leads to the question: The results can be corrupted, right?

Another example of congruence is this: There was a bubble in 1901, a bubble in 1929, and a bubble in 1968. We know that calendar

1 is the same as calendar 29. Calendar 68 is the repeat of calendar 12, but from March 1 to December 31, calendar 12 and calendar 1 merge; the dates and the days of the week coincide absolutely in both cases.

Everyone has heard of the Wall Street stock market crash of 1929; it's easy to recognize. But we have overlooked other downturns in the global economy to deal with events of significance to America or to Europe. Thus, we have not drawn up the other years that belong to calendar 1. We remember that the study we are doing is valid on a planetary level and not on a regional or national level; we must remember the point that the calendar is a representation of the movement of our earth around our beautiful green star.

Looking at the diagram we have called the time diagram, the affinities, and similarities come to mind. We see some calendars intertwined with each other and the relative position of some calendars relative to the consistency of others. For example, calendar 28 merges into calendar 5 from January 1 to February 28, and calendar 28 merges into Calendar 6 from March 1 to December 31. The similarities between calendar 28 and calendars 5 and 6 interlace for the period from January 1 to February 28. Calendar 28 is similar to calendar 5, and from March 1 to December 31, calendar 28 is similar to calendar 6. Calendar 28 is identical to itself, and it is near calendar 12 on its right and calendar 16 on its left. Calendars 12 and 16 are leap calendars, just like calendar 28 itself. So calendar 28 relates to five calendars:

- Calendar 28 is identical to itself:

With calendars 5 and 6, it is the embrace.
With calendars 12 and 16, it is the proximity.
If we take the case of schedule 1:

- calendar 1 is identical to itself.
- it is identical to calendar 24 from January 1 to February 28; it is also identical to calendar 12 from March 1 to December 31.
- it is near calendar 2 on its right, and near calendar 6 on its left.

Also, in this case, we have a group of five related calendars. In this diagram of time, leap calendars constitute the chain threads and non-leap calendars constitute the weft threads in a way. Thus, a calendar, whether leap or non-leap, belongs to a group of five whose relative position with respect to each other is invariable despite movements through the different centuries. The idea that there is affinity within this group comes naturally to mind.

Application

Ms. Goulet was born in 1938. 38 – 28 = 10. Ms. Goulet belongs to calendar 10. There is an interlacing between calendar 10 and calendar 16 from January 1 to February 28; on the other hand, there is an interlacing between calendar 10 and calendar 4 from March 1 to December 31. Calendar 10 approximates calendar 5 on its right, and calendar 10 approximates calendar 9 on its left. Ms. Goulet can therefore have affinities and beneficial relationships with people born during the following years:

(a)  Identity: 10, 21, 27, 38, 49, 55, 66, 77, 83, 94.
(b)  Interlacing:

-    $\Sigma$ from January 1 to February 28: 16, 44, 72, 2000.
-    $\Sigma$ from March 1 to December 31: 4, 32, 60, 88.

(c)  Proximity:

-    Right: 5, 11, 22, 33, 39, 50, 61, 67, 78, 89, 95.
-    Left: 9, 15, 26, 37, 43, 54, 65, 71, 82, 93, 99.

We simply announced the years. If we say 43, it means 1943, and if we say 78, it means 1978. There is a movement of calendars through the centuries; however, there are fixed benchmarks which are the days of the week. These fixed points influence the specific properties of calendars. Calendars move through the centuries with a periodicity of four hundred years; each century has a series of clues characteristic of that century.

# History

In 45 BC, Emperor Julius Caesar asked an Egyptian astronomer, Sosigenes, to draw up a calendar for him. The astronomer first found that the duration of the tropical year was 365.25 days—in reality being the integration of the Egyptian solar calendar into the Roman Empire. So it was clear that every four years, there would be one more day, and the fourth year that added an extra day was brought to the leap year; that is, the year of 366 days, the last two digits are two sixes, hence the leap word. But the rule of adding one day every four years was not well understood, and there was a mess. However, the rule came to be accepted, and one more day was added every four years. This rule continued to apply until the end of the Middle Ages.

In the sixteenth century, the spring equinox was observed to come rather than expected. Also, Pope Gregory XIII undertook to review the calculation. The revision of the calculation showed that the tropical year was not 365.25 days, but 365.2422 days. This explains the delay in the calendar on the real position of the Earth on the ecliptic around the sun. Pope Gregory XIII, by the *Inter gravissimas* bull of February 24, 1582, decided that the day after October 4, 1582, would be October 15. The rule of adding a day every four years was maintained, but at the end of a century, we add a day or not depending on the number of centuries is divisible by 4 or not.

Let me explain: 1600 is a leap because 16, the number of the century, is divisible by four. 1700 is not a leap because 17, the number for the century, is not divisible by four. 1800 is not a leap because 18, the number for the century, is not divisible by four. 1900 is not a leap because 19 is not divisible by four. 2000 is a leap year because 20 is divisible by four. 2100 is not a leap because 21 is not divisible by 4. So immediately, 2400 is a leap, and 2800 is a leap; the rule is well understood. This way of proceeding allows us to limit the delay accumulated over the centuries. This new calendar which corrects the aberrations of the calendar used since Julius Caesar, the Julian calendar, by adding ten days from the start, noting that we pass from October 4, 1582, to October 15, 1582, special treatment for the end of the centuries.

# How To Use Calendars

We have shown in the course of the study that during a whole century, there were fourteen entities which we called calendars which are renewing themselves all the time. Let us reproduce below the 14 calendars with their ambulations which are the renewals of these calendars.

- 1: 7–18–29–35–46–57–63–74–85–91
- 2: 13–19–30–41–47–58–69–75–86–97
- 3: 14–25–31–42–53–59–70–81–87–98
- 4: 32–60–88
- 5: 11–22–33–39–50–61–67–78–89–95
- 6: 17–23–34–45–51–62–73–79–90
- 8: 36–64–92
- 9: 15–26–37–43–54–65–71–82–93–99
- 10: 21–27–38–49–55–66–77–83–94
- 12: 40–68–96
- 16: 44–72–100 (end of the century)
- 20: 48–76
- 24: 52–80
- 28: 56–84

For example, the year 1973 is the renewal of what calendar?

$$73 - 56 = 17$$

17 is a ramble of calendar 6, so 1973 is the reproduction of calendar 6.

Let's find the day of the week that corresponds to April 24, 1973, using the empirical formula.

$$\frac{24 + 6 + 73 + E\,(73/4)}{7} = \frac{24 + 6 + 73 + 18}{7} = 121/7 = 17 + 2/7$$

So it's a Tuesday. If instead of 73 we use 6, we will have:

$$\frac{24 + 6 + 6 + 1}{7} = 37/7 = 5 + 2/7$$

So it's a Tuesday, so we simplified the calculation using calendar 6 by finding the same day. We could have found calendar 6 as reproduced by 73; 73 does indeed belong to calendar walk 6. See the previous table of the walks of the set of fourteen calendars. If we look at the time diagrams, we see that from January 1 to February 28, calendar 6 merges with the 12. From March 1 to December 31, calendar 6 merges with the 28.

In addition, 6 is next to 1 on the one hand, and 6 is next to 5 on the other. We ask the question: what are the relationships between calendars 6, 12, 28, 1, and 5? There are many scientists like Fritjof Capra, former professor at the University of California-Berkeley, who ask us to always seek the interrelationships present everywhere to form integrative realities, that is, nonexclusive.

Calendar 6 is just one example; we could not have been interested in an event that has a relation to calendar 6 or calendar 12. Time diagrams show us the mathematical relationships present between the fourteen calendars.

# Validity of the Gregorian Calendar

We have seen that the tropical year according to Pope Gregory XIII's calculation was 365.2422 days. The corrections are made by using the leap year rule, and by taking into account the treatment of the end of centuries, the corrections aim to ensure that the calendar dates are the most representative of the reality of the position of Earth on the ecliptic around the sun. In a century, leap years are years whose years are multiples of 4. For example, 1604, 1608, 1612, 1620, etc.; and so 4, 8, 12, 16, and 20 are called years of the century. But we note that the leap year which follows 1596 is the year 1600 on the one hand, and we see that the leap year until after 1696 is the year 1704 because seventeen is not divisible by four. Thus, in a century, there are twenty-four leap years if the end of that century is not a leap year. The number of leap years is twenty-five if the end of this century is a leap year.

We have said that the tropical year is 365.2422 days; per year, we therefore have a fraction of a year which is 0.2422 days. In a century, the correction made to make up for time is:

$$0.2422 \times 100 = 24.22$$

We can therefore draw up the following table:

| Period | Number of days due to leap years | Number of days due to real calendars |
|---|---|---|
| From 1601 to 1700 | 24 | 24, 22 |
| From 1701 to 1800 | 24 | 24, 22 |
| From 1801 to 1900 | 24 | 24, 22 |
| From 1901 to 2000 | 25 | 24, 22 |
| Total | 97 | 96, 88 |

We can calculate the timing delay of the Earth's actual position on the ecliptic around the sun. This delay is equal to 97 – 96.88 days= 0.12 days. So we have, for four hundred years, 0.12 days late – about 2 hours 57 minutes. For one hundred years, we are therefore 0.03 days late. The Gregorian calendar would cause three days of delay if a correction to the Gregorian rule was not made. There was therefore a day of delay every 3,333 years.

To overcome this delay, we propose that the year 3200 after 1600 is not a leap, that the year 6400 after 1600 is not a leap, and that the year 10,000 after 1600 is not a leap. So 1600 + 3200 equal to 4800 is not leap, 1600 + 6400 equal to 8000 is not leap, and 1600 + 10,000 equal to 11,600 is not leap. The Gregorian calendar only corrected the delay accumulated since the Council of Nicaea in 345, whereas since its origin, the delay was thirteen days instead of ten days. The Julian calendar has accumulated a delay of ten days in 1,237 years, given 1,582 – 345 = 1,237. The Gregorian calendar will have accumulated three days of delay in 10,000 years. We have the option of reducing this delay to zero by changing the rule for the years 4800, 8000, and 11600.

# Use of Base 7

To facilitate the calculations, in this part, we will redo the calculations to determine the day of the week corresponding to a date using base 7 instead of base 10.

Why base 7? Because the empirical formula lends itself better to it.

Conversion from base 10 to base 7

Let us take the number $N = 1,728$. We divide by 7 this number in base 10. The remainder is $R1$ with $Q1$ the quotient. We write $R1$. $Q1$ is greater than 7. $Q1 / 7 = Q2 + R2$. We write $R2$ to the left of $R1$. If $Q2$ is greater than 7, we continue the division $Q2 / 7 = Q3 + R3$. We still place $R3$ to the left of $R2$.

We continue the division until $Qn$ is less than 7. Then we put $Rn$ of the remainder before, and $Qn$ is to the left of $Rn$. So we have finished writing the number in base 7.

Example 1:

Convert 1416 base 10 to base 7

$1416/7 = 202 + 2/7$ $Q1 = 202$ and $R1 = 2$ so we write 2.

$202/7 = 28 + 6/7$. The remainder is 6. $Q2 = 28$ and $R2 = 6$, so we place the 6 to the left of $R1, R1 = 2$.

$28/7 = Q3 = 4$ is less than 7 and remains $R3 = 0$. We place the 0 to the left of 6 and the 4 to the left of 0. $Q3$ is less than 7 so we stop the divisions, which gives $Q3R3R2R1 = 4062$

Example 2:

Write the number $N = 981$ base 10 in base 7.

$981/7 = 140 + 1/7$, so the remainder is 1, $Q1 = 140$ and $R1 = 1$.

$140/7 = 20$, and the remainder is 0. $Q2 = 20$, $R2 = 0$.

$20/7 = 2 + 6/7$ $Q3 = 2$, $R3 = 6$, $Q3 = 2$ is less than 7 so we stop the divisions.

The number in base 7 is written $Q3R3R2R1 = 2601$ reading from left to right.

## Verification

Let's check the previous example.

$$2601 = 2x73 + 6x72 + 0x71 + 1x70 = 686 + 294 + 0 + 1 = 981$$

Let's take a few examples of dates:

-    The three glorious ones begin on July 27, 1830.

First, calculate the corresponding day of the week with base 10

$$\frac{27 + 1 + 30 + 7}{7} = 9 + 2/7$$

So it's a Tuesday.
Now let's calculate with base 7.
27 is written 36 in base 7. So the calculations are as follows:

$$36 + 1 + 42 + 10 = ?$$

In base 7:

$$36$$
$$+$$
$$01$$
$$+$$
$$42$$
$$+$$
$$10$$
$$122$$

The far-right number in the addition is 2, so it's a Tuesday. In base 7, we just add and read the number furthest to the right. In our case, it's a 2, so it's a Tuesday.

Take another example, July 4, 1776.

$$4 + 3 + 136 + 25 = ?$$

$$004$$
$$+$$
$$003$$
$$+$$
$$136$$
$$+$$
$$025$$
$$204$$

The far-right number is 4, so it's a Thursday.
Let's check this calculation with base 10.

$\frac{4 + 3 + 76 + 19}{7} = 14 + \frac{4}{7}$; the rest is 4/7, so it's a Thursday.

February 26, 1802, corresponds to the birth of the great French writer, Victor Hugo.

Let us calculate with the base 10, the day which corresponds to this date.

$$\frac{26 + 5 + 02 + 0}{7} = = 4 + 5/7$$

The rest is 5/7, so it's a Friday.
Now let's calculate this date with the base 7:

$$35 + 5 + 2 + 0$$

$$35$$
$$+$$
$$05$$
$$+$$
$$02$$
$$+$$
$$00$$
$$45$$

The number furthest to the right is a 5 so a Friday—July 14, 1789, the French Revolution.

$\frac{14 + 3 + 5 + 1}{7} = 3 + 2/7$, so it's a Tuesday.
Let's do the same calculation in base 7:

20 + 3 + 5 + 1 = 32, the rightmost digit is 2 so it's a Tuesday.

# Conclusion

We have established the time diagram for the different centuries, and for each century, March 1 of a calendar always happens on the same day of the week in all the years of a century that belong to the calendar in question. For example, from 1901 to 2000, March 1 of calendar 16, and for the years generated by calendar 16, always happens on a Wednesday. It is the same for calendar 5 and for the years generated by this calendar. If we look at the time chart from 1801 to 1900, March 1 of calendars 5 and 16 is on Friday. What then happens to the specifics or properties inherent in the fourteen calendars? The assertion is true that calendars change in specificity or in ownership over the centuries.

The calendars vary according to the centuries; the days of the week are fixed and do not vary. We could then have the idea that the specifics or properties are determined by the days of the week; we may think that the properties or specifics of calendars are the products of a couple of interactions between the days of the week and the given calendar.

Now that the calendars are known, we can, by going through the archives of the history of mankind, know the product of the days of the calendar week. For example, during a cycle of four hundred years, we can know, without the help of a computer and a calculator, all the calendars where January 1 is a Tuesday, March 1 is a Tuesday, and so on. Thus, we can study the product of a couple of days of the week in a calendar and compile statistics. From the seventeenth century to the twentieth century, the calendars for which March 1 is a Tuesday are as follows: 4, 16, 12, 8, and 4 again. We can do the same to know the product of the interaction of the couple on a given day of the week with the calendars that pass through it over the centuries.

Historians and the archives of mankind can help greatly because historians and archives can tell us what is the human nature of the days of the week and the human nature of the fourteen entities that we limit ourselves to calling calendar. If there is a correlation that appears throughout history, between the days of the week and the calendars, then we can decide where we can extract the properties of the fourteen calendars. Obvious similarities will emerge; the amplitudes of the phenomena can be very variable.

Let us not forget the tremendous influences on our world of parallel worlds about which we know next to nothing. Without getting into esotericism or astrology, we can use the results of our study to decode the realities of the moment. We ask the reader to note that the calendars from 1601 to 1700 are absolutely identical to the calendars from 2001 to 2100 and that the calendar 2013 is absolutely similar to the calendar 2002. I invite all historians, all farmers, all manipulators of media, and all the meteorologists to come and try to decode with me the trends of what will be, the fabric of our realities to come.

In reality, we are in an unknown land. The relationship between the calendars themselves, the improbable interlacing—5 and 28 and 10 and 4, shows that we need a great boost from serendipity because we have often found without seeking what we have long sought without finding.

# Appendices

## Gregory–Kamara Calendars

We have seen previously that calendars 1, 7, 18, 29, 35, 46, 57, 63, 74, 85, 91, and 2002 are all identical: they are the wanderings of calendar 1. We can therefore replace any calendar of this set with calendar 1 to simplify the calculations.

Calendars 2, 13, 19, 30, 41, 47, 58, 69, 75, 86, and 97 are all identical and are the ambulation of calendar 2. So we can replace any calendar in this set with calendar 2 to simplify the calculations.

We also saw previously that the calendar periodicity is twenty-eight years.

Thus, we can replace the calendars in a century with another calendar in the first twenty-eight years for the purpose of simplification. Thus, the 1947 calendar can be replaced by the 1902 calendar because it belongs to its ambulation. The calendar of the year 1970 can be replaced by calendar 3 which belongs to its ambulation. It is easier to calculate with 3 than with 70.

We are therefore going to establish a correspondence between the Gregorian calendars of the century (1, 2, 3, 4, 5, 6, 7, 8, 9, etc.) and the Kamara calendars resulting from their ambulation. The 1950 Gregorian calendar corresponds to the Kamara 5 calendar because calendar 50 belongs to the ambulation of calendar 5 which we will call the Kamara calendar.

The following table is a summary of the correspondence between the Gregorian calendars and the Kamara calendars from 1901 to 2000.

| Gregory | Kamara |
|---|---|
| 1 | 1 |
| 2 | 2 |
| 3 | 3 |
| 4 | 4 |
| 5 | 5 |
| 6 | 6 |
| 7 | 1 |
| 8 | 8 |
| 9 | 9 |
| 10 | 10 |
| 11 | 5 |
| 12 | 12 |
| 13 | 2 |
| 14 | 3 |
| 15 | 9 |
| 16 | 16 |
| 17 | 6 |
| 18 | 1 |
| 19 | 2 |
| 20 | 20 |
| 21 | 10 |
| 22 | 5 |
| 23 | 6 |
| 24 | 24 |
| 25 | 3 |
| 26 | 9 |
| 27 | 10 |
| 28 | 28 |
| 29 | 1 |
| 30 | 2 |
| 31 | 3 |
| 32 | 4 |
| 33 | 5 |
| 34 | 6 |

| 35 | 1 |
|----|----|
| 36 | 8 |
| 37 | 9 |
| 38 | 10 |
| 39 | 5 |
| 40 | 12 |
| 41 | 2 |
| 42 | 3 |
| 43 | 9 |
| 44 | 16 |
| 45 | 6 |
| 46 | 1 |
| 47 | 2 |
| 48 | 20 |
| 49 | 10 |
| 50 | 5 |
| 51 | 6 |
| 52 | 24 |
| 53 | 3 |
| 54 | 9 |
| 55 | 10 |
| 56 | 28 |
| 57 | 1 |
| 58 | 2 |
| 59 | 3 |
| 60 | 4 |
| 61 | 5 |
| 62 | 6 |
| 63 | 1 |
| 64 | 8 |
| 65 | 9 |
| 66 | 10 |
| 67 | 5 |
| 68 | 12 |
| 69 | 2 |

| 70 | 3 |
|-----|-----|
| 71 | 9 |
| 72 | 16 |
| 73 | 6 |
| 74 | 1 |
| 75 | 2 |
| 76 | 20 |
| 77 | 10 |
| 78 | 5 |
| 79 | 6 |
| 80 | 24 |
| 81 | 3 |
| 82 | 9 |
| 83 | 10 |
| 84 | 28 |
| 85 | 1 |
| 86 | 2 |
| 87 | 3 |
| 88 | 4 |
| 89 | 5 |
| 90 | 6 |
| 91 | 1 |
| 92 | 8 |
| 93 | 9 |
| 94 | 10 |
| 95 | 5 |
| 96 | 12 |
| 97 | 2 |
| 98 | 3 |
| 99 | 9 |
| 100 | 16 |

## Links Between Leap and Non-Leap Calendars

Always with the aim of simplifying the calculations, as we have seen in the time diagram, there is a fixed correspondence, whatever the centuries, between the leap and non-leap calendars according to the dates.

Leap calendars consist of two parts: January 1 to February 29 and March 1 to December 31. Leap calendars belong to the spectra of non-leap calendars according to the two periods: from January 1 to February 29 and from March 1 to December 31. (Review the chapter "The Calendars.")

These correspondences are fixed and always remain the same regardless of the century. They are indicated by the following tables:

### January 1 to February 29

| Non-leap calendars | Leap calendars |
| --- | --- |
| 1 | 24 |
| 2 | 8 |
| 3 | 20 |
| 5 | 28 |
| 6 | 12 |
| 9 | 4 |
| 10 | 16 |

### March 1 to December 31

| Non-leap calendars | Leap calendars |
| --- | --- |
| 1 | 12 |
| 2 | 24 |
| 3 | 8 |
| 5 | 16 |
| 6 | 28 |
| 9 | 20 |
| 10 | 4 |

The two previous tables can be represented in another form by the following table:

**Correspondence between leap and non-leap calendars from 2001 to 2100. These correspondences are valid whatever the century**

| Period | Monday | Tuesday | Wednesday | Thursday | Friday | Saturday | Sunday |
|---|---|---|---|---|---|---|---|
| 1er March -31 Dec | 4 | 16 | 28 | 12 | 24 | 8 | 20 |
| 1er January -28 Feb | 4 | 16 | 28 | 12 | 24 | 8 | 20 |

**Month indices : 622  503  514  624**

The previous table reproduces all the calendars of the twenty-first century with a new light from Kamara to simplify the calculations.

The correspondences in this table are valid for all centuries. They can be applied to the twentieth century.

The correspondence table between leap and non-leap calendars combined with the Gregory–Kamara correspondence table allows for quick calculations.

For example, June 6, 1944, corresponds to what day?

$$\frac{6 + 4 + 44 + 11}{7} = \frac{65}{7} = 9 + \frac{2}{7}$$

So it's a Tuesday.

Let's see the light brought by the Kamara calendars.

The 1944 Gregorian calendar corresponds to the Kamara 16 calendar.

According to the table of links between leap and non-leap calendars, calendar 16 is equivalent to calendar 5 from March 1 to December 31.

June 6, 1944, is therefore calculated as follows:

$$\frac{6 + 4 + 5 + 1}{7} = \frac{16}{77} = 2 + \frac{2}{7}$$

June 6, 1944, is a Tuesday.

# TIME DIAGRAMS

# Time diagram   from 1501 to 1600

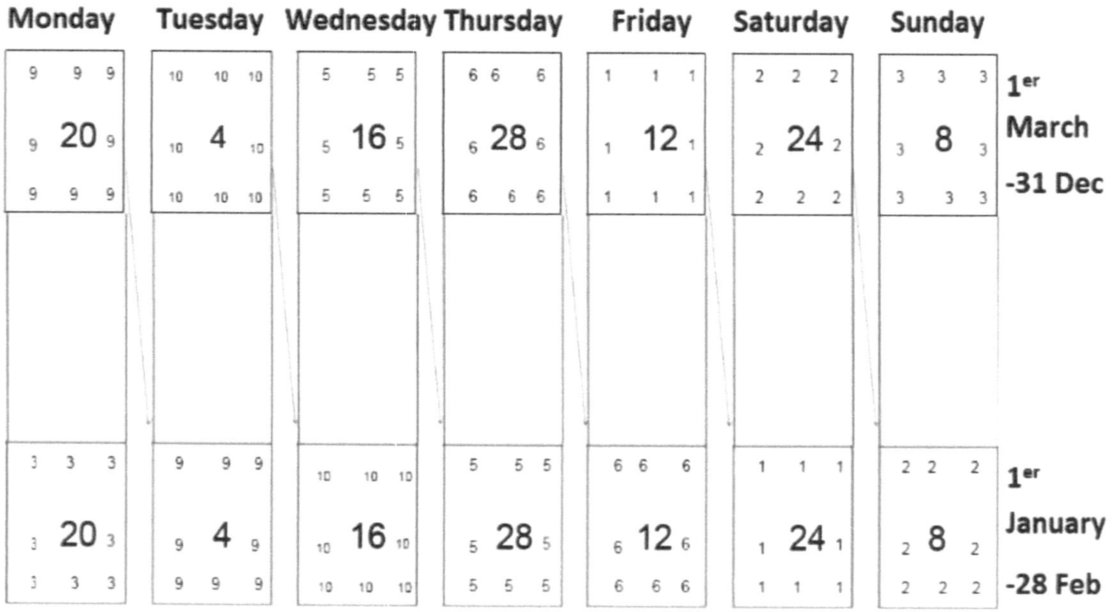

Month indices : 033  614  625  035

# Time diagram   from 1601 to 1700

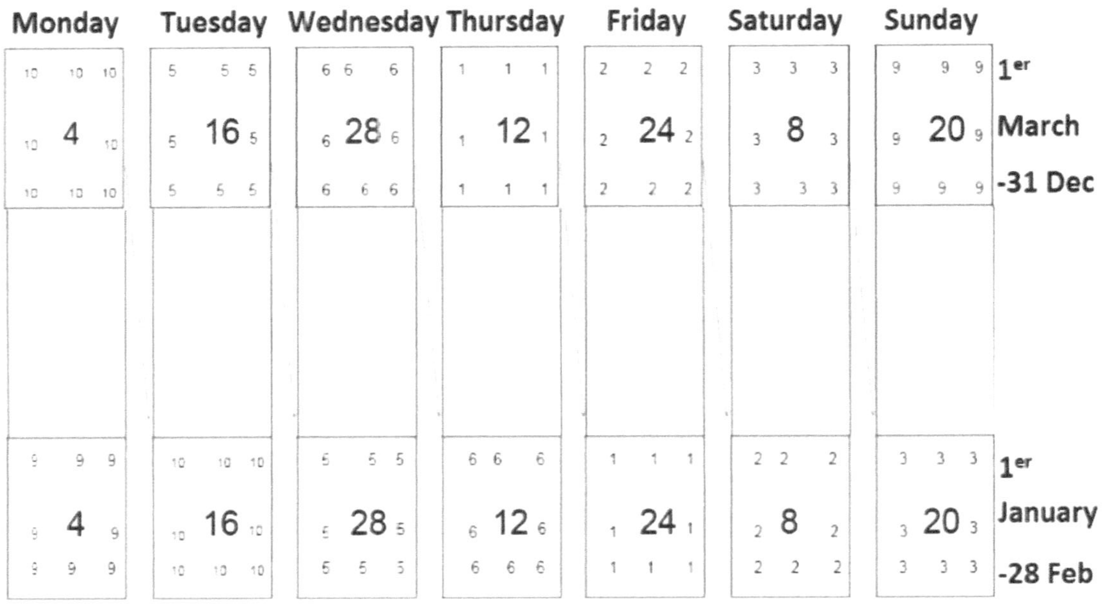


Month indices : 622  503  514  624

# Time diagram from 1701 to 1800

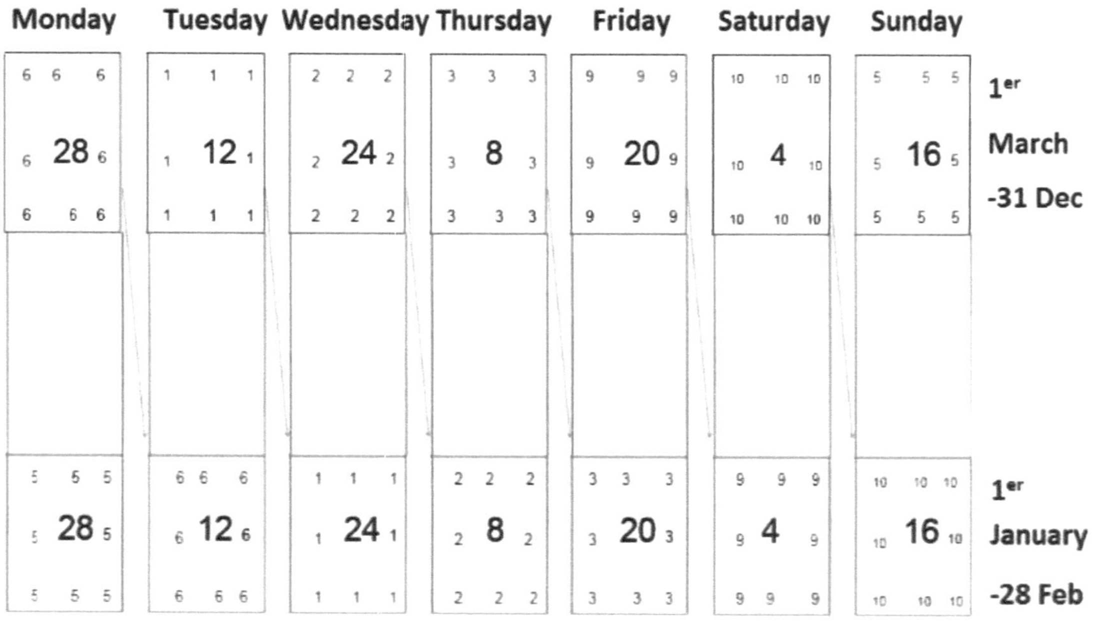


Month indices : 400 351 362 402

# Time diagram   from 1801 to 1900

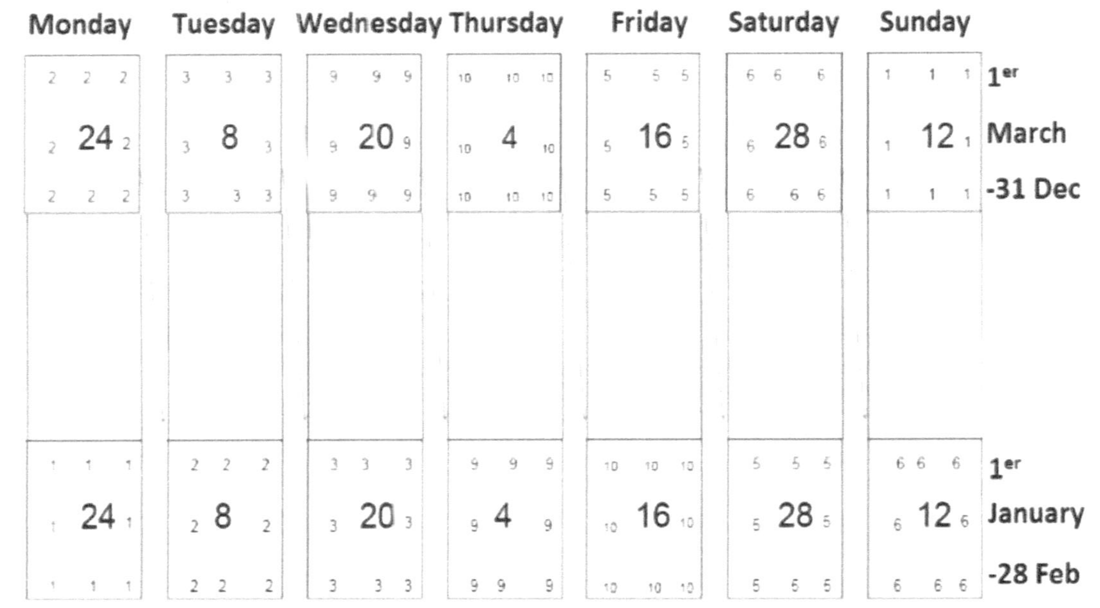



Month indices : 255  136  140  250

# Time diagram from 1901 to 2000

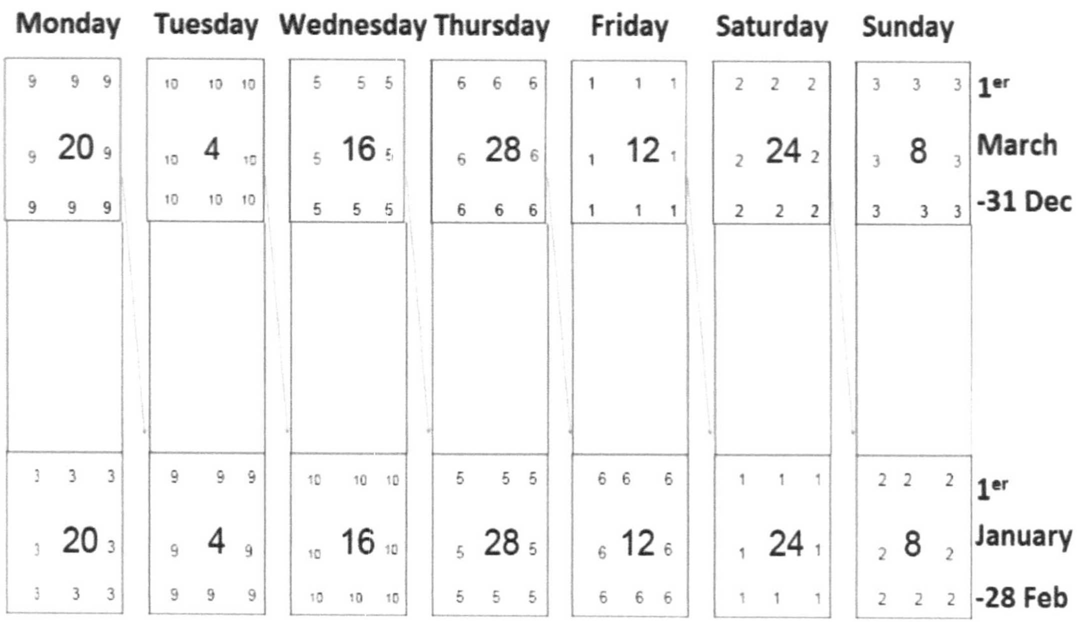


Month indices : 033 614 625 035

# Time diagram   from 2001 to 2100

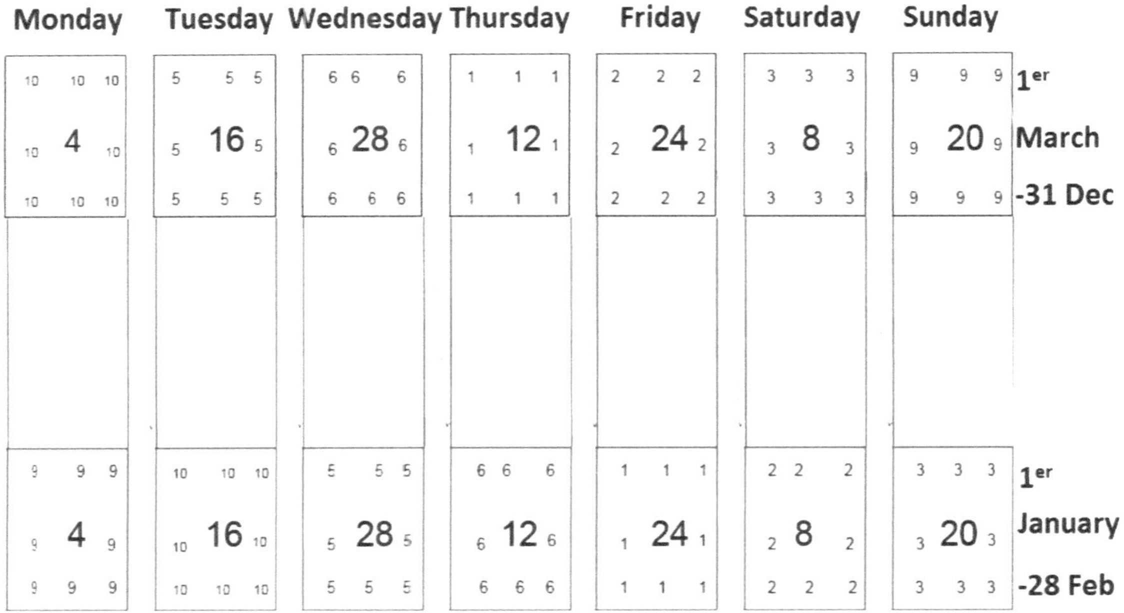

| Monday | Tuesday | Wednesday | Thursday | Friday | Saturday | Sunday | |
|---|---|---|---|---|---|---|---|
| 4 | 16 | 28 | 12 | 24 | 8 | 20 | 1er March -31 Dec |
| 4 | 16 | 28 | 12 | 24 | 8 | 20 | 1er January -28 Feb |

Month indices : 622  503  514  624

# About the Author

Abdoul Koudouss Kamara is a Mauritanian mining engineer born in 1938 who lives in the United States. After completing his primary and secondary studies in Mauritania, he left Senegal to join the Van Vollenhoven High School, now named Lamine Gueye High School.

Afterward he went to France for studies at the University of Grenoble in the MGP section (general mathematics and physics). Then he passes the entrance test of the National School of Mines of Ales in France before returning to Mauritania to occupy positions of responsibility.

In Mauritania, he became the head of the Department of Classified Establishments and Fuels at the Office of Mines of the Ministry of Mines and Industry. After, he held the position of design and test engineer for the company MINFERMA (iron mines of Mauritania), which later became SNIMM (National Industrial and Mining Company of Mauritania). To finish, he occupied the head of security of the same company.

His research in the prediction of events by calendar periodicities dated back to the 1970s. Having lost his sight in 1990, his work greatly slowed down.

Abdoul K. Kamara is also a practitioner of esotericism.